CO_2 Emissions Trading Put to Test

Umwelt- und Ressourcenökonomik

Herausgegeben von

Prof. Dr. Wolfgang Pfaffenberger

und

Prof. Dr. Wolfgang Ströbele

Band 18

LIT

Bernhard Hillebrand
Alexander Smajgl
Wolfgang Ströbele
Jean-Marc Behringer
Bernd Heins
Eric-Christian Meyer

CO_2 Emissions Trading Put to Test
Design Problems of the EU Proposal
for an Emissions Trading System in Europe

Research Results of a Project Team
Headed by
Bernd Heins

Translation by
Valerie Böhner, Sven Flakowski,
Celia Moira Weber, Alexander Smajgl

LIT

Bibliographic information published by Die Deutsche Bibliothek
Die Deutsche Bibliothek lists this publication in the Deutsche Nationalbibliografie; detailed bibliographic data are available in the Internet at http://dnb.ddb.de.

ISBN 3-8258-6654-8

© LIT VERLAG Münster – Hamburg – London   2002
Grevener Str./Fresnostr. 2   48159 Münster
Tel. 0251-23 50 91   Fax 0251-23 19 72
e-Mail: lit@lit-verlag.de   http://www.lit-verlag.de

Distributed in North America by:

Transaction Publishers
New Brunswick (U.S.A.) and London (U.K.)

Transaction Publishers   Tel.: (732) 445 - 2280
Rutgers University   Fax: (732) 445 - 3138
35 Berrue Circle   for orders (U. S. only):
Piscataway, NJ 08854   toll free (888) 999 - 6778

Executive Summary

Emissions trading scheme for CO_2 put to test

Conclusion: Refusal of the EU-Proposal for its conceptional defects and serious negative effects

1. **To take measures against global warming is the task of international climate politics in order to prevent future harms.** The negotiations on UN level, with the help of the Kyoto Protocol and the Marrakech Accords, created the present general conditions. This holds true for the definition of goals and for the Implementation of instruments to attain the set goals. The **European Union (EU)** as a whole as well as its single member states have joined this process of climate policy by accepting a commitment to reduce their Greenhouse Gas (GHG) emissions.

   The German engagement has been clarified in the Burden Sharing Agreement (BSA) of the EU. Germany agreed to a commitment of reduction of 21% in comparison to the base year 1990, which makes for ¾ of the EU wide commitment. The German commitment becomes especially obvious in looking at the reduction of 231 Mt (18.9%) reached since 1990.

2. **The aim of this research project is the analysis of the proposal of the EU Commission from 23.10.2001** to introduce an Emissions Trading scheme for energy intensive sectors and generators on company level.

   An Emissions Trading scheme has to meet two essential requirements:

   - The **conceptional compatibility** with existing international agreements and **practicability** with political, legal, and economic structures.

   - The forecasts have to confirm positive effects in terms of Sustainable Development, which includes the best possible reduction of global GHG emissions with simultaneous consideration of the economic and social effects.

The analysis has to be based on models which enable the comparison of the concrete EU concept as well as modified concepts within different scenarios.

3. The EU proposal contains serious conceptual problems and contradictions:

- The base of the EU concept for an Emissions Trading scheme is the definition of absolute caps on the level of installations. It does not become obvious how the initial allocation can be realized without discriminating installations, as one of the basics of the allocation is the BSA which incorporates very different reduction commitments for each member state. Additionally, early actions which have to be considered for each company since the Kyoto base year 1990 vary strongly.

- Instruments defined on the UN level are excluded (JI: Joint Implementation; CDM: Clean Development mechanism) or restricted (Emissions Trading only between the energy intensive sectors within the EU instead of an Annex B-wide ET). Thus the Emissions Trading scheme is isolated from the Annex B-wide defined instruments.[1]

- If nations were to use the Kyoto mechanisms, this would change the remaining internal split up of reduction commitments for the enterprises within the EU trading system. Such a restriction of agreed upon international law does not make sense.

- It is unclear how the EU proposal can be harmonized with existing national and international instruments of climate change policy without affecting double burdens and high bureaucratic costs.

---

[1] Even the Danish proposal from 28.08.2002 denotes JI and CDM only as „desirable", and the participation of companies at an Annex B-wide ET scheme is not defined.

- Climate change goals have to be implemented in the whole system of political objectives. The EU proposal neglects other goals as energy supply guarantee, growth, and social justice.

4. **With the help of different scenarios this analysis shows that the EU proposal does not fulfill the requirements of a Sustainable Development.** In comparison with modified concepts it results in

    - **higher leakage effects**, which lead to **lower reductions** of global emissions, including a negative effect on existing climate change instruments.

    - **serious negative economic effects** as it evokes the strongest limitation of growth, the worst effect on competitiveness, and the highest losses in production.

    - **significant social faults** due to the highest loss of employment.

5. **Due to these three defects the EU proposal has to be refused.** As the rejection incorporates the ecologic dimension this conclusion does not mean a rejection of climate change policy in general but of the concrete EU concept. The commitments have to be fulfilled by the use of instruments which can be realized conceptionally and which are promising concerning the dimensions of Sustainable Development.

6. **The risk of leakage is a central aspect** resulting from the division of the world in two parts, one with and one without GHG reduction commitments. Every measure in Annex B-regions to reduce CO_2 emissions creates an incentive for production to move from Annex B-countries to Non–Annex B-countries. In fact, CO_2 emissions will be exported and produced goods will be imported. While there is nearly no positive effect on global emissions, a drop in employment will be caused within energy intensive production (base chemicals, cement, glass, paper, iron and steel, ...) and within coal production and coal use. Additionally, significant changes in the national energy mix will be caused. The neglect of the social and economic dimension in connection

with the inadequate attainment of ecologic goals is not compatible with Sustainable Development.

7. **An EU concept has to start from the Kyoto mechanisms in consideration of the existing political, legal and economic systems.** This research project comes to the conclusion that a concept of absolute caps on level of installations cannot be realized due to the conceptional problems it causes. The Danish compromise from August 2002 realizes the need of instrumental flexibility to improve all three dimensions of Sustainable Development but the conceptual defects remain.

To sum up, one can say that the EU proposal must be rejected for its conceptional defects and its unjustifiable ecologic, economic and social effects; in particular due to the bad effects on existing instruments of climate policy. A target-orientated instrument has to fulfill both criteria, the conceptional practicability and the promising forecast concerning Sustainability. In this connection, Emissions Trading can be an efficient instrument.

It has to be examined how an Emissions Trading scheme can be conceptualised in order to fulfill the two criteria. This will be the next step of the research project.

Preface of the editors

On October 23rd 2001 the European Commission and the European Parliament presented a proposal for establishing a scheme for greenhouse gas emission allowance trading within the European Community. At the core of the proposal lies an environmental instrument, which is the trading of certificates. This instrument should be implemented on a legal basis in all Member States at company level from the year 2005 on. Yet, it is not intended that all dischargers of greenhouse gases participate in the trading scheme, instead, trade is restricted to specific installations, to the European territory, and to the greenhouse gas CO_2. Furthermore, the flexible instuments of the Kyoto Protocoll shall not yet be applied to it.

As this proposal might have considerable consequences on the German industry the Mining, Chemical and Energy Industrial Union (IG BCE) initiated a research project that was supported by the companies E.ON, RWE, Vattenfall Europe, Degussa and BASF, and the German steel, hard coal, cement, glass and paper industry associations. The AGEP-Münster (Prof. Dr. W. Ströbele, Dr. A. Smajgl, and E.-Ch. Meyer) and the RWI Essen (B. Hillebrand, J.-M. Behringer) were commissioned to carry out this study. The project coordination and management was delegated to Dr. B. Heins, who acts as the editor of this report as well.

Emissions trading is an instrument that was already agreed upon in Kyoto in 1997 at the state level. But the very special tight trading system according to the EU proposal has to be judged differently. The following analysis shows: The EU proposal is conceptually weak, it is not coordinated with the three Kyoto instruments, and it has negative economic as well as negative ecological effects.

Münster, October 2002

"CO_2 Emissions Trading put to test
– Design problems of the EU draft directive concerning an Emissions Trading scheme"

Contents

EXECUTIVE SUMMARY   V

PREFACE OF THE EDITORS   IX

CONTENTS   X

INDEX OF TABLES   XV

1 INTRODUCTION   1

2 CLIMATE POLICY AND INTERNATIONAL LAW   5
2.1 GREENHOUSE GASES AND REASONS FOR CLIMATE POLICY   5
2.2 THE KYOTO PROTOCOL AND ITS REALISATION   13
2.3 EMISSIONS TRADING IN THE TRANSITION FROM ITS THEORETICAL CONCEPTS TO ITS REALISATION   21

3 RULES OF EUROPEAN CLIMATE POLICY   27
3.1 THE EU BURDEN SHARING AGREEMENT AND ITS PRESENT ACHIEVEMENT OF OBJECTIVES   27
3.2 PREVIOUS INSTRUMENTS OF CLIMATE POLICY IN THE EU MEMBER STATES   29
3.3 THE EU PROPOSAL FROM OCTOBER 23RD 2001   34
3.3.1 THE SUGGESTED REGULATIONS AND THEIR EXPLANATIONS   34
3.3.2 INNER CONTRADICTIONS   38
3.3.3 DECISION PROCEDURE   44
3.4 EU DIRECTIVE AND ENERGY POLICY   45

4 ASSESSMENT OF EFFECTS OF THE EU DRAFT DIRECTIVE   48
4.1 GENERAL ASPECTS   48
4.2 EU ASSESSMENT OF CO_2 PRICES   51
4.3 CRITICAL APPRAISAL OF THE EU ASSESSMENTS   52

5 MODEL ANALYSIS AND COMPARISON OF SCENARIOS 57

5.1	MODEL STRUCTURE *CODE* TM	57
5.2	REFERENCE CASE	57
5.3	SPECIAL CASE GERMANY	61
5.4	SCENARIO STRUCTURE	63
5.5	SCENARIO *EU PROP STRICT*	65
5.6	SCENARIO *EU PROP WIDE*	71
5.7	SCENARIO *MARRAKECH*	73
5.8	STRATEGIC SUPPLY BEHAVIOUR OF HOT AIR	76
5.9	CONCLUDED MODEL RESULTS	83
5.10	MODEL VARIATIONS AND THE INTERPRETATION OF RESULTS	86
5.10.1	THE SECTORAL AGGREGATION	86
5.10.2	CO_2 EMISSIONS CAUSED BY RAW MATERIALS	87
5.10.3	SYSTEM EFFICIENCY OF CO-ORDINATED PRODUCTION PLANTS	88

6 SECTORAL AND MACRO ECONOMIC IMPACTS OF THE EU PROPOSAL 90

6.1	PRELIMINARY REMARKS	90
6.2	FRAMEWORK AND EXOGENOUS DEFAULTS	92
6.3	IMPACT ON THE ENERGY SECTOR	97
6.3.1	SHORT-TERM EFFECTS	98
6.3.2	LONG-TERM EFFECTS	99
6.3.3	COMBINED HEAT AND POWER CYCLE	112
6.4	SECTORAL IMPACTS	114
6.5	MACRO ECONOMIC EFFECTS	127

7 CONCLUSION 130

7.1	THE EU DRAFT DIRECTIVE FROM 23. OCTOBER 2001	130
7.2	THE DANISH COMPROMISE PROPOSAL FROM 28. AUGUST 2002	132
7.3	NET RESULT	133

LITERATURE 135

ANNEX 1: CODE TM – THE APPLIED GENERAL EQUILIBRIUM MODEL 139

MODEL AGGREGATION	**139**
THE MODEL STRUCTURE	**140**
MODELING FOSSIL FUELS	**142**

ANNEX 2: THE RWI MODEL SYSTEM — 145

MODEL STRUCTURE	**145**
ENERGY MODEL	**145**
STRUCTURAL MODEL	**146**
ENVIRONMENTAL MODEL	**147**
SIMULATION POSSIBILITIES	**147**

GLOSSARY — 148

Index of Figures

FIGURE 1: REDUCTION TARGETS ACCORDING TO THE KYOTO PROTOCOL AND THE MARRAKESH ACCORDS ... 20

FIGURE 2: SCHEMATIC PRESENTATION OF THE EFFECTS OF UNIFORM STANDARDS, TAXES AND CERTIFICATES ... 22

FIGURE 3: REDUCTION COMMITMENTS ACCORDING TO KYOTO AND MARRAKESH AND THE EMISSIONS OF 1998. ... 28

FIGURE 4: INFLUENCE OF KNOWN DATA ON THE REFERENCE PATH ... 57

FIGURE 5: MARKET FOR CO_2 EMISSION PERMITS IN THE YEAR 2012. ... 60

FIGURE 6: CO_2 REDUCTIONS AND PERMIT PRICES IF THE EU INTERNAL ET SYSTEM IS ISOLATED AND JI AND CDM PROJECTS ARE NOT ALLOWED ... 66

FIGURE 7: SHARE OF FOSSIL FUELS IN THE FOSSIL FOUNDED ELECTRICITY PRODUCTION OF GERMANY. ... 69

FIGURE 8: CO_2 REDUCTIONS AND PERMIT PRICES WITH AN ISOLATED EU-WIDE SECTORAL ET SCHEME AND WITH JI AND CDM PROJECTS ... 72

FIGURE 9: SHARE OF FOSSIL FUELS IN THE FOSSIL FOUNDED ELECTRICITY PRODUCTION OF GERMANY ... 73

FIGURE 10: CO_2-REDUCTIONS AND PERMIT PRICES UNDER ISOLATION OF THE EU-WIDE SECTORAL EMISIONS TRADING AND WITH JI/CDM-PROJECTS ... 74

FIGURE 11: SHARE OF FOSSIL FUELS IN THE FOSSIL BASED ELECTRICITY PRODUCTION OF GERMANY ... 76

FIGURE 12: CO_2 REDUCTIONS AND PERMIT PRICES IN AN ISOLATED EU INTERNAL SECTORAL SCHEME WITHOUT JI AND CDM PROJECTS, WITH A STRATEGIC SUPPLY BEHAVIOUR OF THE REGION FSU ... 78

FIGURE 13: SHARE OF FOSSIL FUELS IN THE FOSSIL FOUNDED ELECTRICITY PRODUCTION OF GERMANY ... 79

FIGURE 14: CO_2 REDUCTIONS AND PERMIT PRICES IN AN ISOLATED EU INTERNAL SECTORAL SCHEME WITH JI AND CDM PROJECTS, AND WITH A STRATEGIC SUPPLY BEHAVIOUR OF THE REGION FSU 80

FIGURE 15: SHARE OF FOSSIL FUELS IN THE FOSSIL FOUNDED ELECTRICITY PRODUCTION OF GERMANY 81

FIGURE 16: CO_2 REDUCTIONS AND PERMIT PRICES IN AN ANNEX B-WIDE ET SCHEME WITH JI AND CDM PROJECTS, WITH A STRATEGIC SUPPLY BEHAVIOUR OF THE REGION FSU 82

FIGURE 17: SHARE OF FOSSIL FUELS IN THE FOSSIL FOUNDED ELECTRICITY PRODUCTION OF GERMANY 82

Index of Tables

TABLE 1: CO_2 EMISSION FACTORS REFERRING TO FUELS AND GENERATED ELECTRICITY	11
TABLE 2: ENDOGENOUS RESULTING GROWTH RATES OF GDP BETWEEN 1995 AND 2012	58
TABLE 3: EMISSIONS IN REFERENCE CASE IN COMPARISON WITH MARRAKECH/BSA COMMITMENTS	59
TABLE 4: RESULTS WITHOUT A STRATEGIC SUPPLY BEHAVIOUR OF THE REGION FSU	84
TABLE 5: RESULTS WITH A STRATEGIC SUPPLY BEHAVIOUR OF THE REGION FSU	85
TABLE 6: PRICE PATH[1] OF FOSSIL FUELS IN GERMANY, 2000 TO 2020	94
TABLE 7: NOMINAL PRICE PATH OF IMPORTANT PRIMARY ENERGIES IN THE EU, 2000 TO 2020	96
TABLE 8: SHORT TERM REDUCTION COSTS IN THE ELECTRICITY PRODUCTION, €/T CO_2 2010	100
TABLE 9: SPECIFIC FOSSIL FUEL INPUT IN POWER PLANTS, 2005 TO 2020	101
TABLE 10: SPECIFIC INVESTMENT COSTS OF NEW POWER PLANT CONSTRUCTIONS, 2005 TO 2020 IN €/KW	103
TABLE 11: LONG-TERM PRODUCTION COSTS OF NEW CONSTRUCTED POWER PLANTS, 2005 TO 2020, IN €/KWH	105
TABLE 12: LONG-TERM MARGINAL REDUCTION COSTS IN POWER GENERATION, 2005 TO 2020, IN €/T CO_2	106
TABLE 13: LONG-TERM SUBSTITUTION POTENTIALS IN POWER GENERATION[1], 2010 AND 2020, IN GW	107
TABLE 14: SUBSTITUTIONAL CO_2 REDUCTIONS IN POWER GENERATION, 2010 TO 2020, MT CO_2	108

TABLE 15: SUBSTITUTIONAL COST IMPULSES IN POWER GENERATION, 2010 TO 2020, MIO. € 110

TABLE 16: ET AND COSTS OF POWER GENERATION, 2010 TO 2020, €/MWH 111

TABLE 17: LONG-TERM PRODUCTION COSTS IN NEW CONSTRUCTED COMBINED HEAT POWER CYCLES, 2005 TO 2020, €/MWH 114

TABLE 18: SPECIFIC CO_2 EMISSIONS FOR EACH SECTOR AND MEMBER STATE 1999, T CO_2/1000 € GROSS PRODUCTION 117

TABLE 19: LONG-TERM REDUCTION COSTS IN THE CEMENT PRODUCTION, 2005 TO 2020 118

TABLE 20: ET AND ADDITIONAL COSTS IN THE STEEL PRODUCTION, 2010 TO 2020, MIO. € 120

TABLE 21: ET AND ADDITIONAL COSTS OF THE SECTOR GLASS, CERAMICS, AND STONES AND EARTHS, 2010 TO 2020, MIO. € 121

TABLE 22: ET AND ESTIMATED ADDITIONAL COSTS OF THE CEMENT PRODUCTION, 2010, MIO. € 123

TABLE 23: ET AND ADDITIONAL COSTS OF THE NON-IRON METAL PRODUCTION, 2010 AND 2020, MIO. € 124

TABLE 24: ET AND ADDITIONAL COSTS OF THE SECTOR PULP AND PAPER, 2010 AND 2020, MIO. € 125

TABLE 25: ET AND ADDITIONAL COSTS OF THE CHEMICAL PRODUCTION, 2010 AND 2020, MIO. € 126

TABLE 26: MACRO ECONOMIC EFFECTS OF ET, 2010 TO 2020 129

1 Introduction

Measures against global warming are tasks of the international climate policy as precautionary measures for the future. The negotiations at UN-level concerning climate policies have created together with the Kyoto Protocol and the Marrakesh Accords the current conditions. This applies to the ascertainment of the aims as well as to the instrumental implementation to reach the targets. The European Union (EU) as a whole as well as the individual Member States have joined the process and have assumed an overall reduction commitment for the EU as well as individual state commitments. To reach these goals, the EU Commission and the European Parliament presented a proposal for establishing a scheme for greenhouse gas emission allowance trading within the European Community on October 23rd, 2001. Therein an environmental instrument, namely the trading of certificates for greenhouse gases at the company level has been suggested by the year 2005, which should be introduced bindingly into all EU Member States. However, not all dischargers should participate in the emissions trading, only those who run certain installations. Trade should only be valid in the EU and should only include CO_2. Furthermore, the flexible mechanisms of the Kyoto Protocol should not (yet) apply to it.

Since the beginning of the nineties, a multitude of measures have been taken in Germany, with which the emissions of trace gases related to climate change have been reduced clearly and whose continuation will reduce the emissions of such gases in future. The large number of measures reaches from legal regulations and prohibitions over additional taxes and charges to voluntary agreements. This multitude of measures raises the questions if and if so under which conditions the CO_2 emissions trading scheme will be compatible with the already operating national climate program, how extensive will be the additional expected reduction contribution of the CO_2 emissions trading scheme and which sectoral and macroeconomic effects are connected with the implementation of a trading scheme.

The dimensions of possible CO_2 prices, which have been presented by the EU Commission in the explanatory memorandum of the proposal, and with it the foreseeable and far reaching changes of the energy supply and the energy prices in Germany have caused concern about the effects of this very special instrument. Therefore, the Mining, Chemical and Energy Industrial Union (IG BCE), the companies E.ON, RWE, Vattenfall Europe (for HEW, LAUBAG, VEAG), Degussa and BASF as well as the "Wirtschaftsvereinigung Stahl"

"CO₂ Emissions Trading put to test
– Design problems of the EU draft directive concerning an Emissions Trading scheme"

(economic organization steel), the German hard coal mining association and the German paper-, glass- and cement-industry association have ordered a study to analyse the inner logic as well as the calculable consequences of the EU proposal for the energy industry and the energy-intensive industries in more detail. The AGEP-Münster under the leadership of Prof. Dr. Wolfgang Ströbele and the Rhine-Westphalia Institute for Economic Research conducted by the head of the energy research group Dipl. Volksw. Bernhard Hillebrand as a project partner have received the order to carry out this study.

With the following study the project partners present the results of their analysis. In detail the proceeding was as follows:

The conceptual weak points and the inner contradictions of the EU proposal to current international arrangements were brought out in detail.

The consequences for the German and European economy were pointed out which result from the prices for emission permits expected by the EU Commission,

the currently not perceptible co-ordination with the already implemented and approved instruments of the German climate protection policy was discussed,

alternative models of action were developed which take the current valid and international arrangements, in particular the flexible Kyoto instruments, into consideration.

In this respect the study at hand can not only be viewed as a critical analysis of the EU proposal, but also as a contribution to the further development of the existing range of measures of national and international climate policies. The international arrangements concerning climate policies, in particular the Kyoto Protocol and the Marrakesh Accords, are the basis of the analysis. On this basis an arrangement as target-oriented and efficient as possible is being searched. Judged by this claim, the EU proposal was proved critical, in the course of which, above all, the following facts have to be taken into consideration:

Position of the proposal to the Kyoto mechanisms which are international law already;

reduction obligations of the companies and installations involved in the trade and their integration into the national and European reduction targets respectively;

the relation of the companies that are involved in the trade to other emitting companies in the EU and in the Member States;

unsolved conflicts between national and European institutions in the proposal;

possible adverse effects on national climate policy instruments;

transaction and supervision costs of this special trade-model.

According to this goal, the study is structured as follows:

At the beginning, the essential features and their framework of international law will be presented in detail. Here one has to keep in mind that neither the climate problem itself nor the effects of different instruments on already formed reduction goals can neither be quantified nor evaluated without the use of formal methods of analysis and models. Afterwards, the current valid international arrangements, particularly their objectives, but also the agreements concerning the three important instrumental groups Emissions Trading, Joint-Implementation (JI) and Clean-Development Mechanism (CDM), which represent the current possible maximum of international co-operation, will be pointed out. Against this setting, the standard concepts recommended in the literature of environmental economics have to be modified and examined with regard to their practical use. This occurs in the following paragraph.

Based on these foundations the regulations and previous arrangements will be discussed which have been taken at the European level and which belong to the valid institutional and legal conditions that, in the opinion of the EU Commission, will maintain their validity in an Emissions Trading system. In Germany and in other Member States legally determined instruments and proposals to reduce CO_2 emissions and the other five climate-relevant gases belong to the basic conditions. The proposal of the EU Commission to introduce a CO_2 emissions trading scheme is analysed in this context, it will be examined with regard to its conceptual capacity, and the problems that derive from it will be discussed. Goal of this paragraph is to analyse in detail the compatibility of the proposal with the current international arrangements and its transmission in real political, legal and economic structures.

The ecological, economic and social implications of the proposal will be determined with the help of model simulations. Main parameters for these analysis are the reduction goals as well as the necessary reduction measures including the prices for CO_2 emission rights, which result from the reduction measures. First of all the effects will be pointed out which can arise from the dimension of CO_2 certificate prices which have been estimated by the EU itself. The competitive effects, which can be expected, together with the relocation of production and the structural problems of adaptation are of great importance to the global emission balance. The European Union could not calculate these so-called "leakage effects" with the model used. As these effects have consequences for the supply and demand for CO_2 emission rights, lower prices than the initially supposed CO_2 prices in an order of 30 Euro/t CO_2 will arise, however, not only until the process of structural adjustment including the possibly relocation of production has set in. Although the relocation of energy-intensive production does not help to reduce the global greenhouse gas emissions, the EU goals become reachable because of EU CO_2 emissions that will statistically be lower and induce lower CO_2 prices.

These implications will be contrasted with own model calculations in which the international arrangements are explicitly taken into consideration and the flexible instruments of the Kyoto Protocol are integrated into the trade system. The simulation calculations show that not only the negative economic and social consequences of the EU proposal can be avoided with such a broad concept of a trade model, but they show, above all, that the global emissions can be reduced clearly, even if the reduction successes within the EU turn out to be lower than by the implementation of the EU proposal.

Based on these simulations the sectoral and macroeconomic effects of the EU proposal will be examined in an own paragraph. Here the national perspective is the centre of attention; nevertheless, the effects on other Member States will be considered. It is important to have an eye on the other Member States, in particular, if changes of the relative price positions between the Member States result from different production and energy supply structures, which for their part can cause competitive effects within the EU as well as towards third countries.

The analysis closes with a comprising evaluation in which the defects of the EU proposal concerning its conceptual problems and the avoidable sectoral and macroeconomic disavowals of a trade system according to the presentation of the EU Commission are described again. In connection with these results the compromise proposal of the Danish presidency from August 28th 2002 can be analysed critical as in this proposal the until then missing integration of the Kyoto mechanisms is indeed recognised as a serious mistake, but in the remaining explanations no improvement is observable.

2 Climate policy and international law

2.1 Greenhouse gases and reasons for climate policy

Within the environmental policy the so-called greenhouse gases represent a special challenge. On the one hand, they are globally effective, thus they require a world-wide institutional framework. On the other hand, all kinds of damages caused by climate change are not directly observable neither at the present time nor in the past. Different from historical observable damages caused by emissions of direct damaging substances, such as sulphur dioxide or heavy metals, the negative effects of greenhouse gas emissions can only be calculated analytically i. e. by the use of complex models. Insofar climate policy is confronted with a unique environmental decision situation.

a) Anticipatory climate policy only with the aid of model calculations

For millions of years a natural concentration of the so-called greenhouse gases (GHG) has made sure that climatic conditions on earth were such that the life of plants and animals became possible. In addition, the balance of solar irradiation and heat emission is influenced by numerous further factors, such as volcanic eruptions and areal distribution of the ocean current and the rainfall, whose interactions can only be evaluated with the aid of complex models. As a consequence of human activities, additional parameters have been of interest to scientists for about 10 – 15 years. Firstly, it is a fact that fossil fuels, which combustion makes an important contribution to the total emissions of GHGs, came into use for commercial purposes only after the year 1709.[2] Other GHGs like FCKWs are clearly of later origin. In addition, only the development of information technologies laid the foundation to estimate the effects of anthropogenic emissions, which can be expected on a fairly long term basis on the world climate at all. In this sense, it is a hardly known phenomenon which can have far-reaching ecological, economic and social consequences. At present, at least about 90 % of the commercial energy supply of mankind is based on the use of fossil energy resources, by their combustion CO_2 is released.

The concentration of the so-called greenhouse gases in the atmosphere is in turn subjected to diverse natural processes of composition and decomposition.

---

[2] Abraham Darby built the first furnace based on hard coal in Wales in the year 1709. This invention meant the turn away from charcoal and is regarded to be the beginning of the use of fossil fuels for industrial production.

The reproduction of this processes in the form of mathematical equations proves to be a very complex task which could hardly be solved without an efficient information technology. Not until these preconditions are fulfilled, it is possible to reproduce the natural processes reliably and calculate fairly long-term predictions for the foreseeable concentrations in 50 or 100 years. In particular only an extensive knowledge of, for example, the so-called carbon cycle allows to quantify effects which lead from human activities and influence the world climate, for the lion's share of 95 % of the global CO_2 production belongs to natural sources. Even the insignificant fuzziness of specifications in environmental models can have consequences for the basis of the estimation of anthropogenic influences. There is also the fact that natural climate fluctuations can be traced back more than several million years, human influences, however, have been recorded only for 150 years. As a result of this, it is difficult for methodical reasons to capture the importance of the anthropogenic influences quantitatively exact.[3]

Despite these restrictions, there is consensus that complex model systems are indispensable for the analysis of climate change. This also applies to the estimation of the consequences of changed GHG concentrations concerning the climate, form, amount and areal distribution of the rainfall, ocean current, transmissions of vegetation zones, etc.[4] In this context, of course, one must consider that GHG emissions result not only from running human activities, such as the combustion of fossil energy sources, the cultivation of wet rice, animal husbandry or the release of FCKWs', but also from natural processes, such as volcanic eruptions or the release of methane gas out of permafrost grounds.

Comprehension with regard to these processes, which were acquired by means of such model calculations in the nineties, led to the view that international efforts to reduce manmade GHG emissions are necessary to prevent a long term change of the world climate, which is sometimes strikingly called climate catastrophe. From the economic point of view this is a great challenge as the composition of the atmosphere is a public good, which means that nobody can be excluded from using this good and there is nonrivalness in consumption,

---

[3] Thus it is not astonishing that there are doubts about the accuracy of the statement that a climate catastrophe caused by humans is approaching. See for example Neumann-Mahlkau (2002), pp. 2.

[4] As a rule this modelling is carried out by meteorologists. In Germany beside numerous others Professor Graßl and Hasselmann from the Max-Planck Institute of meteorology in Hamburg are the leading scientists with international influence. Neumann-Mahlkau (2002) shows, for example, that there are conflicting opinions concerning anthropogenic causes.

whose previous use occurred without any restrictions: The extraction of oxygen and the emission of gases, such as CO_2 or other GHGs, have taken place, free of charge, for hundreds of years. Isolated efforts of individuals or groups to pursue climate policy can hardly help. A co-ordinated world-wide proceeding of all relevant emitters would be most appropriate.

In terms of a cost-benefit analysis especially the US-American scientific research has tried within the last eight to ten years to evaluate the benefit as well as the costs that could arise if mankind has to adjust to a changed climate. If the climate models forecast a clear rise of the sea level, the costs that could arise will outweigh the benefits although the costs have been corrected further down in the model calculations several times. For it is not at all à priori certain that the effects caused by the relocation of vegetation zones or the changing of rainfall must only be negative. From the environmental point of view, the possible consequences are nevertheless assessed rather negative as the number and the dimension of damages has risen because of the increasing frequency of storms and floods.[5]

For the estimation of ecological, economic and social consequences of an efficient climate policy, the use of environmental as well as economic models is indispensable. This applies particularly to the analysis of effectiveness of potential tools, such as special emissions trading systems. With recourse to this formal systems, potential environmental effects in the form of emission reductions can be estimated as well as economic effects at the level of sectoral and national economic possibilities of growth and development under diverse policy conditions.[6] The last point serves as a basis for the estimation of the social dimension of the sustainability concept. However, in the face of the complexity and unsolved issues of the institutional organisation of policy, these models can only serve as first approximations of possible effects with regard to if-then statements. As a rule, in particular, there is a "trade off" between the wanted detailing of single processes and the calculability of such models. These models would possibly be overtaxed if they should be used for the immediate answering of very specific questions, for example, with regard to the effects on indivudual companies and processes of production. For the mainly used model types include either a relative detailed reproduction of single production tech-

---

[5] In Germany the Potsdam Institute for Climate Impact Research conducted by Prof. Dr. Schellnhuber is the leading institute in this field.

[6] This work refers exemplary to the works within the context of the German project IKARUS and to the general equilibrium model **code-TM**.

nologies that can be compressed to sectoral production sectors by aggregation or they form the sectoral production processes with the aid of nested production-functions (nested CES-functions). Both assessments have very specific advantages and disadvantages. Their results have to be analysed against the setting of the chosen approach.[7]

This introductory presentation of the problem structure points out that there are doubts about both the possible damages caused by the undamped continuation of certain GHG emissions and the implications of reduction strategies. Therefore, political decisions should consider the possibility of an error in both directions.[8] In this context, measures to limit greenhouse gas emissions in view of provisions should be introduced. Despite of the uncertainty concerning the possible environmental and economic implications, it is indisputable that formal methods are indispensable in many fields of climate policy. However, for the interpretation of the results the contexts and mechanisms which are reproduced in the prevailing models as well as the limits of the analysis which emerge from these mechanisms and contexts have to be considered. In this way the EU Commission did not consider the following points in their model analysis:

- Relocation of energy-intensive production processes or
- feedback of the emissions trading system on world energy prices.

a) Globally effective greenhouse gases and international climate protection

The changes of the heat emission capacity of the earth are a result of an extremely complex process through accumulation of greenhouse gases in the atmosphere. As this accumulation takes place regardless of where the greenhouse gas is released, possible changes in climate are a global problem. This means that they are independent of the location and with regard to the next decades to a large extent independent of the moment of the emission. Therefore, a solution of this problem requires international arrangements and co-operation. Actors of such international co-operation solutions are primary the nation states.

In the international conferences of Rio de Janeiro (1992) and Berlin (1995) first steps were taken to regard the limitation of the emission of greenhouse

---

[7] That is why the price reagibility of the energy consumption is underestimated by technique-orientated bottom up models and overestimated by AGE models.

gases as a political field of activity for the international community. The United Nations Framework Convention on Climate Change (UNFCCC) was signed in Rio de Janeiro. In appendix I the Industrial Nations were told to limit their GHG emissions. This target was not yet connected with specific obligations and instruments. In the following Conference of Parties (COP1) in Berlin in 1995 the „Berliner Mandat" was passed, in which for the first time the needed obligations of the Industrial Nations concerning a quantified reduction of GHGs were listed as a target for future negotiations.[9]

As the first international treaty the Kyoto Protocol from 1997, which was adopted from the third Conference of Parties in Japanese Kyoto, laid down the needed emission limits with regard to time until 2012, with regard to location for the so-called Annex B Countries (Industrial and Transformation Nations) and with regard to medium for six important greenhouse gases (carbon dioxide, methane, nitrous oxide, hydrofluorcarbons, perfluorcarbons, sulphur hexafluoride). In experts opinion this list comprises the most important GHGs. Single GHGs stay for a different period of time in the atmosphere and have divergent greenhouse effects, which can be converted into CO_2 equivalents with the aid of certain calculation processes. In this way, they can be standardised.[10] Even if the present discussion concentrates, above all, on carbon dioxide, the other GHGs should not be ignored. They are of special interest just because of their different relative "ease" of emission reductions or avoidances.

---

[8] McKibbin/Wilcoxen emphasise, for example, that such a care is suitable, (2002), pp. 107.

[9] At the Berlin conference the former German Chancellor Kohl intensified the already formulated German reduction target for CO_2 by changing the basis year from 1987 to 1990 without agreement with the functional departments. This specific German CO_2 target for the year 2005 does not have a binding effect under international law.

[10] This is defined as the specific greenhouse potential that puts the radiation effect of a gas over a specific period of time in relation to the same amount of the reference gas CO_2. The so-called global warming potential (GWP) of a greenhouse gas is defined as the ratio of the time-integrated radiative forcing from the instantaneous release of 1 kg of a trace substance relative to that of 1 kg of a reference gas. Direct radiative effects occur when the gas itself is a greenhouse gas. Indirect radiative forcing occurs when chemical transformations involving the original gas produces a gas or gases that are greenhouse gases, or when a gas influences other radiatively important processes, such as the atmospheric lifetimes of other gases. If a retention period of 100 years is supposed, the resulting factors are 21 for CH_4, 310 for N_2O, 23900 for SF_6. These factors oscillate for PFC at around 6500 (perfluoromethane) and 9200 (perfluoroethane), for HFC at around 140 (HFC-152a) and 11700 (HFC-23). For details see Bundesministerium für Umwelt, Naturschutz und Reaktorsicherheit (ed.), Klimaschutz in Deutschland – Zweiter Bericht der Bundesrepublik Deutschland nach dem Rahmenübereinkommen der Vereinten Nationen über Klimaänderungen, Bonn 1997, pp. 82.

b) The "pollutant" CO_2

Carbon dioxide shows the greatest share of the six greenhouse gases; this climate gas reached in the year 2000 about 80 % of the total emissions of GHGs in the EU.[11] CO_2 differs from the other classic pollutants sulphur dioxide, nitrogen oxide or dust as it is not directly harmful. However, it is in broad limits even essential for the growing of plants. These limits are extremely rare exceeded, for example, just in case of an accident,. Therefore, it serves as a basis for every life on earth. The harmfulness results merely from the accumulation of CO_2 in the earth's atmosphere and the deduction of changes of the earth's heat balance through climate models.

Up to now, CO_2-emissions are not measured directly. They are determined indirectly with regard to the carbon content of the used fuels. The fuels itself are investigated in the context of exaltations and are subsumed in terms of energy balances. Up to now the necessary reports of the companies have been without negative economic consequences for the reporting firms so that they can be classified as reliable. However, in reality fossil energy sources are never completely homogenous. There is natural gas with a bandwidth of different chemical compositions. The intermixture and the upper heating value of hard coal and lignite is scattered, too. The mineral oils are also of different composition according to their country of origin which is expressed, for example, in diverse sulphur contents and viscosities.

On average, the CO_2 emission factors compiled in schedule 1 are valid for the fuels used in Germany in the nineties:

---

[11] The contribution of CO_2 to the accumulated stock in the atmosphere is – measured in CO equivalents – clearly lower, because here, for example, CFCs are also included, whose (additional) emissions are close to zero at least in Europe.

Energy sources	t CO_2 / MWh	Efficiency in the generation of electricity	t CO_2 / MWh electricity
Lignite	0,40	41,0 %	0,97
Hard coal	0,34	42,5 %	0,80
Fuel oil	0,27	40,5 %	0,67
Natural gas	0,20	55,5 %	0,36

Table 1: CO_2 emission factors referring to fuels and generated electricity

The higher the efficiency of the used technology and the lower the carbon content of the used fuels, the lower are the CO_2 emissions under equal conditions. But the differences between the carbon contents of single fuels can only be compensated in narrow limits by improving the efficiency: The highest specific emissions are accounted for lignite. Lower emissions are steadily accounted for hard coal and less than the half to natural gas.

Carbon dioxide is a specific GHG because of several aspects:

- There is no extensive retention technology at reasonable costs so that a filter technology analogous to the realisation of the German "Großfeuerungsanlagen-Verordnung (GFAVO)" for sulphur and nitrogen oxide as a relative simple solution is ruled out. Thus experiences of the GFAVO or other concepts of classic pollutants are not useful. A trade with SO_2-emissions in the Middle West/North East of the USA (SO_2-ATP) is only partly possible. The retention technology for CO_2, which is in discussion for major emitters, is just at the beginning of its development.

- Up to now, a reduction of CO_2 is only possible by means of an increase of efficiency, that is to say by means of substitution through capital (energy savings with the aid of better heat insulation, better motor technology, ...) or rather by means of substitution of energy sources high in carbon with energy sources poor in carbon or carbon-free. With regard to the carbon content lignite and hard coal are inferior to mineral oil and natural gas; nuclear energy as a carbon free source of energy is superior to all fossil energy sources. The same applies to regenerative energy sources, whereas, over and above that, these sources are in comparison with nuclear energy free of radioactive pollutants and waste products.

"CO₂ Emissions Trading put to test – Design problems of the EU draft directive concerning an Emissions Trading scheme"

- CO_2 is a world-wide operating substance, which as a rule does not cause acute damages. However, it develops its effects only on a long term basis through accumulation in the earth's atmosphere. In this respect CO_2-emissions in 2005 or 2010 are just as equal as emissions in India or Germany. As a result of this a local or a regional political assessment can only help partially solving the problem. A global course of action would be appropriate for such a problem. This applies also to the Kyoto Protocol, in which merely a part of the world signed up for reduction targets.

- There is no "hot spot" problem concerning CO_2. This means there are no local or regional dangers that would make the recourse of legal regulations and bans necessary in case of an acute averting a danger. For that reason strategies of avoidance can be planned as a provision for future dangers so that looking at it from this aspect the use of a trade system is useful.

- In contrast to the classic air pollutants, such as sulphur dioxide or nitrogen, CO_2 does not cause concrete visible damages.[12] In any case the damage effects can only be deduced on the basis of model calculations. Measures to limit damages have to overcome considerable higher acceptance problems than conventional measures to fight against acid rain, to reduce the ozone build up in the lower atmosphere, the less so since the model calculations were revised and specified again and again with regard to the prognosticated rise of the sea level.

- There is a proportional relation conditioned by the law of nature between the use of fossil energy sources and the specific CO_2 emissions. As a rule an elaborate determination of direct emissions is not necessary because these emissions can be determined indirectly on the basis of the used fuels.[13] In practice, however, this requires uniform methods concerning energy balances and recording of different chemical compositions of the primary energy sources.

The close relation between fossil fuels and CO_2 problems signifies that the climate protection policy comes inevitably into contact with the targets and instruments of energy policy. Whereas, it influences the energy economy as

---

[12] GHGs that cause other damages are exempt from this: CFCs for example do not only change the earth's balance of heat, but cause damages already today to the ozone layer, which is especially important as a shield against ultraviolet radiation close to the polar caps.

[13] Of course, in this connection significant incentives arise to smuggle or to camouflage purchases if the dedicated prices of CO_2 or penalties are clearly above the procurement prices of the fuels.

well as energy-intensive sectors with regard to the position of competition. From this point of view, climate policy is at the same time always energy, competition and general economic policy, too.

2.2 The Kyoto Protocol and its realisation

The third Conference of Parties took place in Kyoto in December 1997. In this conference a protocol was passed after difficult negotiations that contains a binding commitment of reduction for the so-called "Annex-B" countries[14] (the Industrial Nations together with East Europe and the former USSR) according to international law. This protocol says that the whole emissions of CO_2 and of five other greenhouse gases have to be reduced about at least 5 %. The European Union (EU) as a whole as well as single Member States have taken over a clearly higher commitment of reduction (8 %).[15] These commitments of reduction will become binding under international law if the protocol is ratified by at least 55 countries and if the emissions of the Annex-B countries, in which the protocol was ratified, amount to at least 55 % of the whole emission of this group of states in the base year. The obligations of reduction are not only related to CO_2, but also to methane (CH_4), nitrous oxide (N_2O), hydrofluorocarbons (HFCs), perfluorocarbons (PFCs) and sulphur hexafluoride (SF_6).[16] The reduction targets do not have to be fulfilled until the year 2005, but can be achieved five to seven years later. Thus, the target is no longer based on an appointed year (for instance 2005) but on a period (2008-2012). Instead of exclusive national reduction options, supranational options can basically be exploited by means of trade of emission permits (article 17, ET), by means of the joint realisation of CO_2 reduction strategies between Industrial Nations (article 6, JI) or between Industrial and Developing Nations (article 12, CDM).

---

[14] Annex B contains the reduction commitments determined in the Kyoto Protocol for individual countries or groups of countries respectively. Although this list coincides by and large with the list of industrial and transition countries in Annex I of the Framework Convention on Climate Change, it is not identical with it. Croatia, Liechtenstein, Monaco and Slovenia are added to the list. Separated from these countries the Czech Republic and Slovakia are listed. Turkey is no longer on the list.

[15] In addition to the reduction commitments determined in the Kyoto Protocol, the EU agreed on a new internal burden sharing agreement in July 1998. This agreement deviates clearly from the commitments of the Kyoto Protocol and imposes a reduction commtitment for Germany of about 21 %.

[16] See United Nations Framework Convention on Climate Change (ed.) (1998), Adoption of the Kyoto Protocol to the United Nations Framework Convention on Climate Change Kyoto, FCCC/CP/7/Add.1, Decision 1/CP.3, Annex A, 18.3.1998, p. 28.

More than four years after the passing of the Protocol the ratification process has not been finished yet. Meanwhile, the United States have retired from this process. Beside to single Member States, the European Union has signed and in the meantime ratified the Kyoto Protocol. For the obligations of the European Union as well as the single Member States as "jointly acting Parties" it is important that according to article 4, section 5 "each Party to that agreement shall be responsible for its own level of emissions set out in the agreement" if the joint reduction target will not be reached. These individual reduction commitments are set to 8 % for all Member States. In contrast, the EU internal burden sharing agreement provides a broad diversity of reduction commitments for individual Member States. Luxembourg (-28 %), Denmark and Germany (each -21 %), Austria (-13 %) and Great Britain (-12,5 %) agreed on the strongest reduction rates, whereas other countries, such as Spain (-13 %), Greece (+25 %) and Portugal (+27 %), are allowed to increase their emissions. If Luxembourg, which is not comparable to other EU Member States because of its size and two special factors steel industry and energy imports, is not taken into consideration, it is obvious that two major emitters, namely Germany and Great Britain, as well as two smaller countries, namely Denmark and Austria, gave far reaching reduction promises, whereas especially the countries bordering the Mediterranean, with the exception of Italy (-6,5 %), are allowed to increase their emissions significantly.

Meanwhile, this Burden Sharing Agreement (BSA) has also a binding status according to international law as it was integrated into the EU ratification of the Kyoto Protocol.

The most important Kyoto agreements at a glance:	
➢ On average greenhouse gas emissions (CO_2, CH_4, N_2O, PFCs, HFCs, SF_6) in the Annex I countries should be at least 5% below 1990 levels in the period 2008/12.	
➢ Reduction of GHG emissions compared to 1990[17]:	
- EU, Liechtenstein, Monaco, Switzerland, Bulgaria, Lithuania, Latvia, Estonia, Romania, Czech Republic, Slovakia, Slovenia	92%
- USA	93%
- Japan, Canada, Hungary, Poland	94%
- Croatia	95%
- Russia, Ukraine, New Zealand	100%
- Norway	101%
- Australia	108%
- Iceland	110%
➢ By 2005 demonstrable progress in achieving climate protection should be made.	
➢ In principle the following flexible mechanisms are possible for an efficient achievement of reduction targets:	
- Emissions Trading at the Annex B level	
- Credits through projects within Annex B countries (JI) or in developing nations (CDM).	
- Substitutions within the group of greenhouse gases	
- Substitution between earlier and later periods *(banking)*	

Scheme 1: The Kyoto agreements at a glance.

The Kyoto Protocol comprises reduction commitments as well as the possibility to reduce the increase of the relevant stocks of GHGs by among others the enhancement of sinks, the promotion of sustainable forms of agriculture, the phasing out of market distortions. For this reason the Kyoto Protocol is wider conceived than generally noticed in public discussions.

---

[17] For some countries, such as Hungary (average of 1985-87), Poland (1988), Bulgaria (1988), Romania (1989) and Slovenia (1986), other base years were determined.

According to article 6, every party listed in Annex B[18] may transfer to, or acquire from, any other party emission reduction units (resulting from projects aimed at reducing anthropogenic emissions by sources or enhancing anthropogenic removals by sinks of greenhouse gases). From an international point of view the use of such projects is denoted as Joint Implementation (JI). The concrete modalities should be defined in later conferences. Joint Implementation is of special importance if within the Annex B countries diverse technical standards are predominant. In Germany, for example, the iron and steel production calculates an energy use of approximately 14 –15 GJ/ t raw iron and the East European iron and steel production in Poland and in the Ukraine respectively calculates 24 GJ/ t raw iron.[19] In this case instead of an elaborated expensive improvement of the German production of a few percentage points, the same expense can help to improve clearly the net emission reduction elsewhere.

Article 12 gives Annex B countries the opportunity to use project activities in non-Annex B countries to contribute to compliance with part of their quantified emission limitation and reduction commitments. Thereby a technology transfer for an advanced energy use from the Industrial Nations to Third World countries should be encouraged. The necessary rules of certification should be developed in later conferences. This mechanism has developed an additional flexibility for the realisation of the commitments and is denoted as Clean Development Mechanism (CDM) in the international discussion. This assessment is of great importance because, on the one hand, the strongest increase of GHG emissions can be forecasted for non-Annex B countries and, on the other hand, there exist great inefficiencies in the energy system and less expensive reduction potentials in the basic industry and industrial goods industry that can be mobilised in case of an induced technology transfer. As an example for the iron and steel industry in China, on an average, an energy intensity of 33 gigajoule (gj) per t raw iron was assessed. This is up to the use of old technologies and an inefficient energy use: After all, the German equivalent is only half as high as the Chinese factor.

---

[18] The Annex terms are to some extent confusing for laymen because in the Framework Convention of Climate Change and, therefore, in the Kyoto Protocol the expression Annex I comes up. But there are actually two individual annexes in the Kyoto Protocol: Annex A and Annex B. Annex A contains definitions of greenhouse gases and production processes. Annex B contains the list of the restriction commitments of the Industrial Nations. The countries which are affected match with two exceptions with those specified in Annex I. If we talk about the definition of GHGs and the like, we refer to Annex A in the following, the national reduction commitments are in Annex B.

[19] See Buttermann (1999), pp. 145.

According to article 17, parties listed in Annex I may participate in emissions trading whereas "any such trading shall be supplemental to domestic actions for the purpose of meeting quantified emission limitation and reduction commitments under Article 3". Thus, Emissions Trading (ET) is not bound to projects, but it creates a market for the use of abatement cost differences. As the Kyoto Protocol regards only contracting parties as relevant actors, it is not clear if this trade initially arranged for countries can or should be relevant for individual companies or even households or small consumers.

The Kyoto Protocol created a framework for climate policy which should be clearly defined at the following conferences in Buenos Aires (COP 4), Bonn (COP 5) and Den Hague (COP 6). As in Den Hague despite of intense negotiations no agreement with regard to a protocol could be reached, the sixth Conference of Parties was adjourned for some months and was continued as COP 6 bis in Bonn in summer 2001.

During this COP 6 *bis*, the concretion of the applicability of the flexible instruments specified in the Kyoto Protocol was almost reached. A last withdrawal of Japan and Russia adjourned the final agreement to the COP 7 in Marrakesh, which took place in October/November 2001.

In Marrakesh[20] an agreement on the rough structure for the application of the flexibility instruments was signed. The diplomatic tug of war between the individual contracting parties led to the so called Marrakesh Accords concerning the fungibility of the emissions trade, the project measures divided into JI and CDM and the CO_2 sinks. Furthermore, the agreement on the temporal dimension of the reduction commitments, the so-called banking, as well as the agreement on the necessary compliance system and the accompanying monitoring were signed.

The trade with emission rights will start in the year 2008 and comprises four different types: *Emission reduction unit* (ERU), *Certified emission reduction* (CER), *Assigned amount unit* (AAU) and *Removal unit* (RMU).

- AAU stands for the reduction quantity that results from the approved reduction commitments according to article 8 of the Kyoto Protocol. Within the Emissions Trading abated amounts can be sold as AAU. In contrast to ERU and CER, AAU can comprise the so-called *hot air*.

---

[20] See UNFCCC (2001).

- ERU serves as a basis for article 6 of the Kyoto Protocol and defines emission reductions which result from carried out JI projects. Reductions that originally belong to indigenous projects are defined as ERU in case of trading and are settled up each time. JI projects will be credited from the year 2008. A crediting of projects carried out previously is not intended.

- Article 12 of the Kyoto Protocol defines CER. CER is limited to emission reductions which are achieved in CDM projects and which will be certified by an CDM-committee at UNFCCC-level. This committee comprises ten members of whom four are from the Annex B countries and six from non-Annex B countries. As the development of sinks for CO_2 falls under CDM, a guideline on the crediting of such measures will be worked out up to the COP9 (2003). Until then, a registration of removals by sinks will not take place. All the other CDM projects can be registered and credited retroactively by January, 1st, 2000.

- RMU presents emission reductions, which have been reached because of the indigenous net-effect of sinks for CO_2, whereas, article 3 paragraph 4 of the Kyoto Protocol serves as a basis. Initially on the basis of the *Bonn Agreement* (COP 6*bis*) only effects concerning forestry were taken into account, effects of land use (on the whole denoted as LULUCF – *land use, land use change and forestry*) were ignored.

Therefore, if a German company carries out an emission reduction project with a Romanian power company, the German company is credited ERUs. If a project on the reduction of CO_2 is carried out with a Chinese steel company, we talk about CERs. Both cases result in effective emission reductions. If already in the run-up the Romanian power company or the Chinese steel company has reduced its production or carried out measures to reduce emissions by own means, these amounts can be offered according to the arrangements of Kyoto and Marrakesh, for example, to the German state in the form of AAUs. RMUs have an effect on the national balance of greenhouse gases and, therefore, influence the target that must be reached at national-level. This target can serve as a basis for a guideline of reduction targets for individual sectors.

This definitional differentiation has become necessary owing to the decision to credit reductions that are achieved by individual flexibility instruments at different levels on the indigenous reduction target. In this way, a new restriction to the emission rights arise. Every party, in principle, which has accepted and ratified the Kyoto Protocol as well as the Marrakesh Accords takes part in the international Emissions Trading. Every party is obliged to hold back a certain quantity of emission rights (CPR: Commitment period reserve) in order to prevent the uncovered selling of emission rights. If a country emits more than designated in the Kyoto Protocol because of too high sales of emission rights, it is not allowed to sell further emission rights until the current GHG commitments (as a minimum limit) are fulfilled.

The banking comprises the possibility to transfer existing emission rights into subsequent periods. While doing this, AAUs can be transferred without restrictions. ERUs and CERs, however, can each only be transferred up to 2,5 % of the whole committed emission quantity. RMUs can not be transferred to subsequent periods.

The Marrakesh Accords had a decisive influence on the charging of sinks for CO_2. As figure 1 shows, these commitments have partly clear effects on the reduction commitments of the Kyoto Protocol. The highest increase was attributed to Russia. Now Russia can by the average of 2008/12 raise its emissions by 4 % compared to 1990. The available potential of *hot air* is de facto clearly higher. Seen absolutely, 120 Mio t CO_2 equivalent are at Russia's disposal, additionally. The Marrakesh Accords make a positive impact on the Canadian situation, too. Canada must no longer reduce 6 %, but is allowed to emit an additional amount of 1,2 % (44 Mill. t CO_2). A similar situation arises for Japan. Japan is allowed to emit almost 48 Mill. t. CO_2 additionally. Beside these three parties, which are typical of considerable modified reduction commitments, the position of the EU has hardly changed. Now the Marrakesh Accords have pined down the position of the EU to a reduction quantity of 7,5 %. A marginal reduction of the absolute reduction commitment (reduction of a maximum of 4,5 Mill. t CO_2 per year by using all sinks) is applied to Germany, whereas the German share of reductions at EU-level has even risen. As a result of this, Germany benefited only infraproportional of the reduced EU burden. Instead of the initially accepted 254 Mill. t CO_2 equivalent according to the Burden Sharing Agreement, Germany must merely reduce 250 Mill. t CO_2 equivalent until the year 2010.[21]

---

[21] See Schafhausen (2002), pp. 90. Additional sinks from Russia, Canada and Japan alone amount to the same order of magnitude as the whole German reduction commitment.

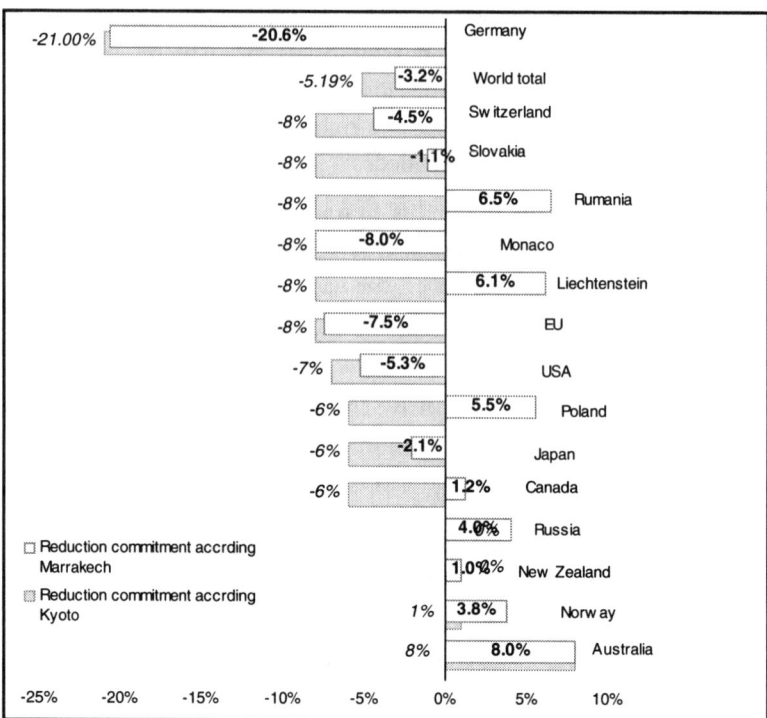

Figure 1: Reduction targets according to the Kyoto Protocol and the Marrakesh Accords[22]

Whereas these arrangements and international agreements, which have been described above, define the procedures of the countries, a new problem arose on the next level. The question is how the accepted commitments and instruments can be transferred to the level of individual economic subjects (companies, state departments as energy consumers, road-users, private households, etc.) so that effective reduction measures are initiated. In this lies the decisive challenge of the environmental policy: In the end, concrete measures, investments of companies and home-owners, behavioural changes as well as migration of companies and the replacement of indigenous energy-intensive production by additional imports decide on the success of climate policy.

---

[22] The crediting of sinks is, in no case, clearly arranged. The data underlying this illustration were investigated by RWI in co-ordination with the BMU.

2.3 Emissions Trading in the transition from its theoretical concepts to its realisation

a) The concept

Economic theory essentially knows three approaches of political interventions and, therefore, of internalisation of up to now external effects, such as the emission of pollutants:

- **Legal regulations** were with regard to GHGs tighten up in Germany within the last years (heat insulation standards, methane release out of dumps, ...). Tax solutions refer to certain final sources of energy (mineral oil, natural gas, electricity), whereas, special provisions exist for reasons of international competitiveness concerning the energy-intensive sectors.

- **Marketable Emission Permits** present a direct quantity rationing. Their theoretical advantages are unquestionable, practical experiences, however, are rather meagre. The few practical cases, such as for SO_2 or NO_x in the US, are not transferable to CO_2.[23] Therefore, the EU proposal for establishing a scheme for greenhouse gas emission trading brings up numerous problems.

- **Co-operative solutions** between government and companies or associations respectively strive for a "voluntary commitment" to reach certain reduction targets. Actually, such approaches work because the involved economic entities fear severe penalties in case they do not reach their individual target.

- **Liability law** is of no importance for the case of GHGs because of the large number of emitters and of the specific damage mechanism.

Beside desired allocative effects, every method implies distributional effects because of the intervention of the state itself which differs, however, with regard to its severity between the sectors and individuals.

---

[23] After the implementation of an SO_2 emission allowance trading system at relatively low costs, Sulphur dioxide emissions in the mid-west of the USA were reduced by changing to coal with a lower sulphur content and the installation of desulphurisation units. The trade was made possible at relatively low bureaucracy costs because a sufficient but yet manageable amount of power plants was involved.

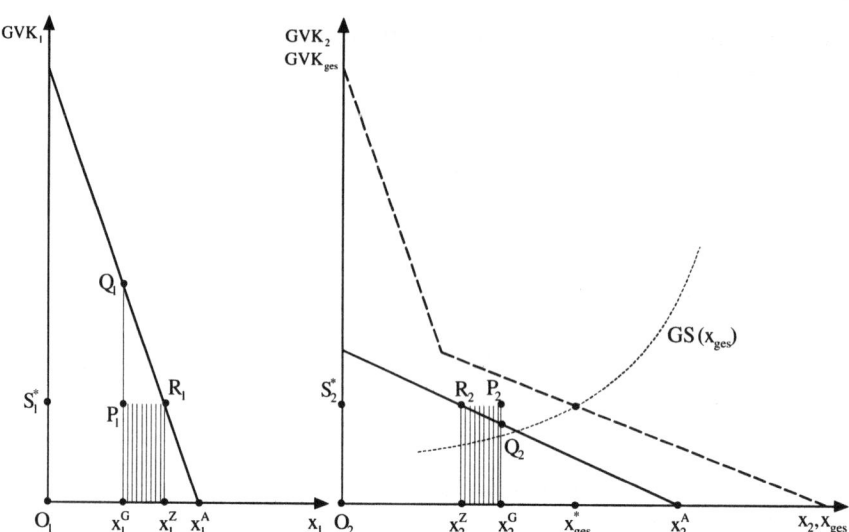

Figure 2: Schematic presentation of the effects of uniform standards, taxes and certificates[24]

If (for analytic purposes) perfect information of the environmental department is assumed vis-à-vis two distinct emitting firms A and B, the following is valid with regard to the instruments:

In an initial position without any environmental policy both firms emit x_i^A, out of which total emissions result from horizontal addition. In order to make this argument clear, we suppose that the marginal damage function is appreciable so that a bisection of the previous total emissions to x_{ges}^* is optimal.

Uniform standards ("everybody reduces 50 %") x_1^G or x_2^G respectively are, as a rule, inefficient: The same total quantity of emissions could be reached if firm 1 – with relative "expensive" abatement costs – reduces its emissions to a level of x_1^Z and firm 2 reduces to the quantity x_2^Z. The saving of costs in case for firm 1 in the amount of $P_1R_1Q_1$ is greater than the additional costs $P_2R_2Q_2$ for firm 2. x_i^Z is defined with regard to the fact that the last abated unit of emissions causes the same (marginal) abatement costs, namely $0S^*$ for every firm.

For the tax or certificate solution without allowances it is true that the effects of a tax with a tax rate $0S^*$ comply exactly with the effects of certificates issued by auctions. The total financial burden for the dischargers amount in both cases to $0S^* R_i x_i^A$ = sum of taxes or certificate costs $0S^* R_i x_i^Z$ plus costs of

---

[24] Similar schematic presentations are in Cansier (1993), chapter 7 or Fees (1995) chapter 4 and 5.

emission abatement (= area of the triangles below the marginal abatement costs up to x_i^A). In each case the optimal levels of emissions are reached.

The alternatives with allowances correspond with each other likewise:

- Certificates with *individual allowances* issued by grandfathering x_i^Z work like taxes with individual allowances x_i^Z. Here every single discharger bears the individually different emission reduction costs on its own (= triangles).

- Certificates with *uniform allowances* (for example as grandfathering with a uniform percentage depreciation → x_i^G) comply in its effects with taxes with uniform allowances. In this case discharger 1 must buy certificates amounting to $x_1^Z - x_2^G$ at a price 0S*, which were offered by discharger 2.

Voluntary commitments, in principle, can reach the same allocative effects as long as the participants act explicitly or implicitly on the assumption that a breach of commitments will be punished. Depending on the intenseness of cooperation between the respective associations and their ability to implement the agreed targets also internally, this instrument can come close to the efficient allocation.

Consequently, the *allocative effects* of taxes and certificates, at a time, with and without allowances (with regard to certificates = grandfathering-share) are in the case of a comparable design identical. The same applies to *voluntary commitments* if they can be implemented internally credible. The actual financial burden for the companies is, however, dependent on the design of the allowances and the methods of issuing costless basic equipment of emission rights respectively. Consequently, if it is sometimes for fear of taxes plead for certificate solutions in political discussions, one is mistaken: Taxes and costless issued emission rights can be equivalent in an appropriate allowance system.

The *advantages* of a system of a more flexible division of individual reduction targets towards uniform standards and, therefore, towards a *certificate system* are big if preferably numerous dischargers with distinct different marginal abatement costs (steepness, position) are included in a pool with internal compensation possibilities. On the contrary, the advantages are very low if the marginal abatement cost curves run quite similar.

However, if there would be, in addition, different emission reduction commitments for individual dischargers or groups of dischargers, defined by politics, which would have to be obeyed separately, this would abolish or restrict the advantages of a flexible certificate system. Such institutional inefficient plans would lead to a divided market: Within this restricted subset with its own commitments other prices would be valid than for the other participants without special restrictions. The problematic nature of the EU proposal from October 23rd 2001 lies in such an additional formation of subsets.

b) The transformation of primary into final sources of energy

Energy sources in their natural shape (mineral oil, lignite and hard coal, wind, water, solar energy, uranium, ...) are practically not directly usable for important applications. Merely natural gas can - possibly after purification[25] or blending to a standardised calorific value - be used directly. Final sources of energy like petrol, fuel oil, electric current or briquettes can only be used for energetic applications after having been converted in suitable plants. The energetic net losses depending on the plant (refinery, power station, ...) are likely to be around 70 % in the case of coal-fired power stations and around 60 % in the case of modern plants. On the use of electricity produced in coal-fired power stations for instance, this means that the total CO_2 emission accrues, for technical reasons, in the plant itself. In the case of a gasoline engine, however, about 95 % of the total CO_2 emission are discharged from cars and only about 5 % from refineries. As a result of this it is necessary to define at first where policy has to tackle the problem.

Therefore, it is conceivable that the climate protection policy should turn its attention to the production or the import of primary sources of energy respectively (**upstream assessment**) or to the combustion of primary and final sources of energy (**downstream assessment**). With respect to an easy collection of data this could be easily answered in favour of the upstream assessment, if, objectively correct, the climate protection policy followed a global uniform assessment. According to the Kyoto Protocol, this spatial range is only valid for a certain section so that the generation of electricity in a country such as Algeria or the production of petrol or fuel oil in Kuwait takes place without restrictions. Thus, the incentive for Annex B countries emerges to import highgrade (final-) sources of energy, such as electricity or mineral oil products, as well as basic substance chemicals from non-Annex B countries. This means that the transformation losses emerge in the non-Annex B countries. Conse-

---

[25] One of the largest sulphur factories in Europe is in Großenkneten close to Oldenburg, where sulphurous natural gas is purified before it is fed into pipelines.

quently, this is ultimately of no use to climate protection: Merely the European CO_2 statistic could score a "success".

c) The realisation

Climate protection policy produces a **global public good**. No state, no company, no farmer or household can be excluded from its benefits. Analysis and predictions given by natural scientists give rise to the assumption that at the earliest in the middle of the century one must reckon the fact that considerable damages caused by anthropogenic GHG accumulations will come into existence.[26] Because of long planning and investment periods in the energy industry, countermeasures in the sense of an anticipative policy should be taken in time.

In contrast to these facts, the costs incurred of reducing GHG emissions have to be beard today by those who reduce their GHG emissions in comparison with a "business-as-usual"-path. There is also the fact that by means of international agreements a limitation of GHG emissions could not be reached for the greater part of the world. According to quite realistic valuations, countries like China and India will discharge comparable orders of magnitude within few years like the United States emitted in the year 1990. In this respect, a high incentive for free-rider behaviour comes automatically into being.[27]

Thereby environmental policy on the basis of the Kyoto Protocol does not have the required conceptual coverage. Consequently, beside the above mentioned possibility to reduce emissions by technical measures, any unit that is located in one of the Annex B countries has the option to migrate to a country where this restriction is not binding. Actually, this so-called leakage effect is highest where energy costs amount an important share of total costs. An industrial company with a relative low share of energy costs will not migrate because of climate policy and the resulting regional increase of prices of energy sources. As soon as the use of fuels is made more expensive and/or electricity costs ascend steeply by means of regional effective measures, this aspect is, however, of great relevance to the energy-intensive basic sector and to the indus-

---

[26] Such statements are under the proviso that major changes, such as the re-routing of the Gulf Stream, are not set off.

[27] Additional energy related CO_2 discharges of China between 1990 and 1999 alone amount to the same order of magnitude as the whole German reduction commitment following the Kyoto Protocol.

trial goods sector as well as to the conversion sector (power stations and refineries).

Political constellations led to the Kyoto Protocol. It is only due to the formation of a subset comprising the Industrial Nations as a group, which submits to GHG commitments, that the international co-operation took place. In this given and real situation the instrumentation of the climate policy plays an important role: All measures should be taken in such way that, in fact, a reduction of global GHG emissions arise.

Statistic "quasi-profits" of nations and groups of nations by means of migration of energy-intensive production into non-Annex B countries, without being useful to the emission reduction target, should possibly not occur. Consequently, beyond the theoretical foundations, all instruments have also to be analysed from the additional point of view that they cause as low as possible leakage effects. Such measures would especially be contra productive if they induce the migration of activities from the EU in other countries which produce with disadvantageous technologies or a less effective energy mix so that on balance even an increase of GHG emissions is induced.

Exactly this additional restriction must be taken into account in case of the implementation of the conceptual basics to a real environmental policy as, otherwise, there is the danger that the particular range of emissions is not properly registered by the particular environmental policy. These arguments do not mean that one can do without climate policy because the Kyoto Protocol installed an institutional inefficiency, but they mean merely that the theoretical concept is not applicable without modification to climate policy in case of the choice of instruments and that an independent analysis is necessary in case of the process of implementation and its instrumentation.

With regard to the EU proposal it is pointed out that just this constraint, namely to provide as little as possible incentives to relocate energy-intensive production, can not be modelled explicitly in the model used by the EU. The leakage problem as an endogenous mechanism does not exist.

3 Rules of European climate policy

3.1 The EU burden sharing agreement and its present achievement of objectives

The negotiations of Kyoto were accompanied by negotiations concerning the EU internal division of the reduction commitment of 8 % altogether. These negotiations led to the so called "burden sharing agreement" between the Member States of the EU. As a result of structural policy considerations each European Member State agreed to individual reduction commitments. Germany and Great Britain accepted the highest reduction commitments. Due to the Marrakesh Accords and the possibility to allow for setting of CO_2 sinks some of the individual reduction targets were diluted but the cumulative change for the EU was rather small. Figure 3 shows the resulting changes compared to the reduction targets of the Kyoto Protocol.

Although part of the reduction commitments could be met by important countries until 1999, the overwhelming majority has to be met in the coming years up to the first budgeting period. In addition to this, one has to take into consideration that reductions reached until 1999 are attributed to some extent to the German reunification (restructuring of the East German economy, introduction of a better technology and fuel-adjustment of the households and small consumers) and to the competitive opening of the British electricity sector (about 50 Mill t CO_2). The reductions are also partly attributed to special developments that will not recur in the following years. In particular, the German successes after the reunification were only made possible by raising considerable capital investments by companies and by the support of all German taxpayers for support of the restructuring processes. In this respect a designation for CO_2 abatements as "wall fall profits" is factually incorrect.

Furthermore, a division into sectors makes clear that the additional reached abatement successes are attributed exclusively to the energy sector and to the industrial sector, whereas the other sectors, in the first place the transportation sector, are responsible for a further increase of emissions.

"CO₂ Emissions Trading put to test
– Design problems of the EU draft directive concerning an Emissions Trading scheme"

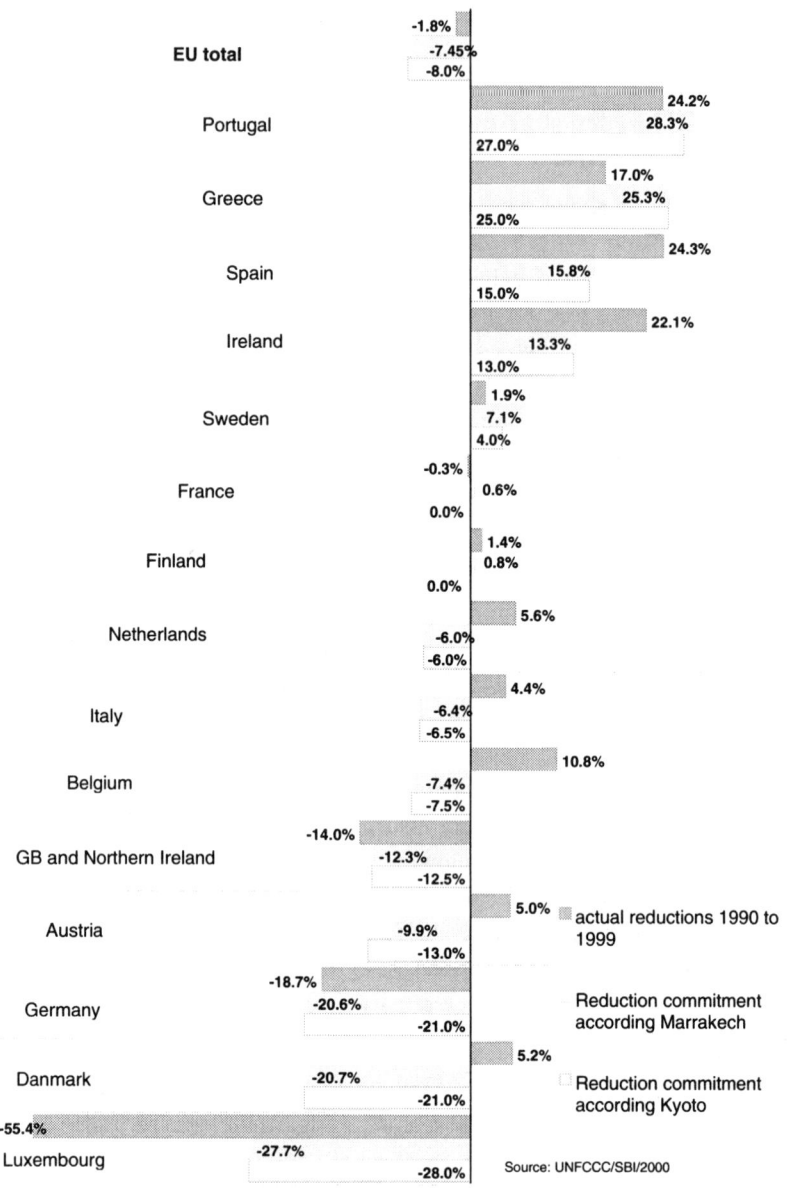

Figure 3: Reduction commitments according to Kyoto and Marrakesh and the emissions of 1998.

3.2 Previous instruments of climate policy in the EU Member States

In the former years the EU Member States have partly implemented very different instruments regarding climate policy. The repertoire reaches from setting legal standards over voluntary commitments to environmental taxes and few approaches concerning the installation of a scheme of tradable permits. In Germany **additional instruments**, such as the climate agreement from November 2000, the combined heat and power (CHP) law from April 2002 and the Renewable Energy Sources Act, have been installed in the energy-intensive sectors and in the conversion sectors (power stations, refineries). Important contributions to abatement that are party legally fixed are expected from this instruments within the next years. Severe standards or tax signals in the form of an environmental tax apply to other fields.

An integration of national climate policies with the flexible instruments of the Kyoto Protocol has not taken place yet. This is based on the fact that rules for the implementation of the Kyoto mechanisms have only been available since the Marrakesh conference. According to the above mentioned explanations of the explanatory statements for the flexible instruments of the Kyoto Protocol, this integration is of fundamental importance.

On the other hand, the EU intends to introduce uniform rules for an emissions trading scheme by means of the EU proposal presented in October 2001. However, these rules imply that an emissions trading starts in 2005, that this trade is only valid for special parts of the industrial sector, namely for the operators of special plants, and that the GHG CO_2 is traded only. In the following these aspects are described in detail.

a) Double effects

The cost and price effects of these legal frameworks already in place are of importance to a European trading scheme as these effects can intersect with the cost and price effects of a rigid trading system. Thus, the Renewable Energy Source Act[28] – similar to the CHP law – aims at improving the competitiveness of electricity generating technologies, which in the middle term can make an important contribution to the abatement of CO_2 emissions.

---

[28] The Renewable Energy Source Act to support the use of regenerative energy sources, such as water, biomass, wind energy a/o.

Thereby it is crucial for a future trading scheme that these financial incentives are implemented. Independent of the precise design of a trading scheme, one can expect that the production costs and therefore the electricity prices will rise due to the liability to acquire emission permits. Although the concrete effects are depending on the structure of the portfolio of generation capacities, on the efficiency of the energy conversion, and on the substitution effects, which are induced by emissions trading, it is obvious that the competitiveness of the generation of electricity on the basis of regenerative sources of energy will become better as a consequence of increasing electricity prices. The improvement of the CHP technology is dependent on the definition of the trade system and its effects on the electricity prices. It is important that the payment rates determined in the CHP can be reduced so that they do not harm the competitiveness of those technologies. Beyond it one can assume that electricity of special regenerative energy sources or of CHP processes will substitute conventionally produced electricity and that the resulting emission abatements or the then redundant emission rights can be sold to the licence market. Here very different positive and negative effects can arise depending on who appears as the operator of the actual plants. If renewable or rather CHP plants and plants that are based on fossil fuels are part of the portfolio of generation capacities of a single operator, a "double dividend" can arise depending on the national method of allocation of emission rights.

Similar double effects are to the greatest possible extent excluded from the environmental tax reform. According to the laws of the commencement and continuation of the environmental tax reform, the use of energy in specific environmentally friendly CHP plants and in very efficient gas and steam turbine plants was liberated from the environmental tax and the present mineral oil tax. In this respect, the use of energy for the generation of electricity and/or heat remains tax-exempt even if this conversion is connected with environmental impacts and with a considerable consumption of energy resources. Therefore, an overlapping of an emissions trading scheme and environmental taxes is not possible in the field of energy conversion.

This does not automatically apply to final consumption because the emissions trading will influence the electricity price level as the additional costs including the revenue of sales of emission abatements in electricity generation can possibly not be completely neutralised by a restructuring of the portfolio of generation capacities and are insofar cost and price effective. In case of an unchanged electricity tax, these price increases have to be added to the direct burden of the ecological tax reform. Consequently, a cumulative effect arises at this level. Without correction of the electricity tax law this effect can have consequences for the competitiveness of energy-intensive production sectors, particularly as in this case the double securing of the environmental tax (restrictions and reduced tax rates) does not take effect.

In this respect the emission trading system can enlarge hitherto unincisive differences of the financial burden between sectors and thereby it can accelerate a structural change over and above CO_2 abatements. As a consequence, indigenous energy-intensive products are replaced with imports from third countries. Thereby the total quantity of GHG emissions would not be reduced world-wide and consequently, a use for the world climate would not arise. These connections have to be taken into consideration adequately in case of a realisation of a trading scheme. Hitherto, it is not foreseeable how the EU proposal will take account of this.

b) Declaration on global warming prevention and CHP law

The declaration on global warming prevention of German Business from March 1996 and its extension by the agreement on climate protection from November 2000 are of paramount importance to the national climate policy. In this agreement the federal government and German Business have agreed on the fact that the German economy will reduce its specific CO_2 emissions about 28 % by the year 2005 and its specific emissions of the other six greenhouse gases about 35 % by the year 2012. In return the federal government will not take any initiative to achieve climate protection through command and control measures and will endeavour to guarantee that on the European level due account will be taken of the contributions made as regards EU-wide CO_2/energy taxation.

In this context a special agreement relating to the support of CO_2 reductions by the use the combined heat and power was reached on June 25th. By this agreement an absolute emission abatement of 10 million t by the year 2005 and of 23 million t by the year 2010 as against 1995 will be reached.

If under this conditions, in addition, the CO_2 emissions discharged by the transportation sector, households and small consumers can be stabilised, Germany will reach its commitment determined in the EU burden sharing agreement (BSA) by the year 2010. Germany has taken visible efforts in order to reach its climate protection targets and it has already reached a large part of its reduction commitments with the help of its hitherto implemented instruments.[29]

Even if both instruments are directly designed for the reduction of emissions and therefore differ fundamentally from additional taxes or charges, the preconditions and effects of the two approaches are thoroughly different. Central to the agreement on climate protection is a **reduction of specific emissions in the industry and in the public generation of electricity**.

A trade system itself does not lead to a reduction of GHG; it should act as an incentive to reduce GHG emission at those installations where a reduction can be reached relatively easy. If thereby the individual commitments are fixed for costless allocated GHG emissions in the form of absolute upper limits for every company or rather every installation, that is to say a so-called cap-and-trade-system is installed, a relocation to non-Annex B countries will be expected beside the indigenous reduction by means of energy saving and fuel substitution. In contrast to this, the polluter pays principle applies to the agreement on climate protection in the strict sense, that is to say emissions result not only from the direct use of carbonaceous energy sources but also from the consumption of energy sources free of carbon, whose generation is connected with CO_2 emissions – therefore, especially electricity and heat.

Thereby it is important to bear in mind that such fuels that are used in industrial plants to generate electricity and heat, are part of the total industrial energy consumption. Consequently, they are also found in the industrial CO_2 balance. Hitherto, up to now energy consumption is balanced on a net basis in order to avoid double counts so that it is only CO_2 effective if it is not generated in own plants. This net consumption is rated with a constant CO_2 factor, practically according to the base year 1990. According to this concept, CO_2 reductions in the industrial sectors can only be reached by reducing the net external electricity consumption or by improving efficiency and substituting fuels in the industrial energy and heat generation. CO_2 reductions that arise from efficiency improvements or fuel substitution in public electricity generation influence only the CO_2 balance of the public supply, not the balance of the industry. This way of setting-off could not be used in case of the EU proposal.

---

[29] Thereupon, calculations of the EU of which Germany is able to mobilise in addition to the accepted commitment further lower-cost reduction potentials in which merely 30 €/t CO_2 are calculated become quite questionable as Germany is hardly able to appear as a supplier for 30 €/t CO_2.

A similar principle applies to such reductions that arise from an increased use of electricity and heat generation from combined heat and power plants in public supply. Only emission reductions resulting from an improvement of efficiency and fuel substitution in production are allowed to be assigned to the public district heat supply, whereby an increase in the power to heat ratio has to be regarded as an improvement of efficiency.

If this concept should also apply in an emissions trading system, the public electricity and district heat supply can only enter the trade with such emissions that arise from the improvement of efficiency and fuel substitution in the field of public supply. Decreasing or increasing CO_2 emissions resulting from changes of electricity and heat consumption may, with regard to the initial allocation and heat consumption, only be assigned to consumers, but not to public generation.

To summarize, the implementation of a trading scheme without consideration of measures already in place can evoke considerable disavowals and incompatibilities which can affect the effectiveness of hitherto measures and instruments as well as the effectiveness of a trading scheme.

The following points are of a particular problematic nature:

- The conceptual shift from a downstream approach to an upstream approach leads to a completely new orientation of climate policy.
- The successful voluntary agreements of German industrial sectors can no longer work in the new regime according to the EU proposal.
- The danger of politically unwanted distribution effects, especially a doubling of burden, affect environmental taxation as well as the rules of the CHP law.
- The compatibility of legal standards, which require the best possible technology (IVU guideline), with an emissions trading system, based on absolute emissions, is unsettled.

c) Flexible instruments in other EU Member States.

In comparison to other EU Member States, above all, Great Britain, Denmark and the Netherlands implemented own emissions trading systems. These are designed very individual.

On the basis of a voluntary participation Great Britain provides an incentive by the exemption from the Climate Change Levy (environmental tax) and by the adoption of abatement costs through an auctioning of reduction commitments. Thereupon, the payment of abatement costs, as in the case of a "normal" emissions trading system, is not made by another company with higher marginal abatement costs (win-win-strategy), but by the taxpayer. As a result of this, on the one hand the abatement costs[30] are reimbursed completely for companies that are major dischargers and on the other hand an extra bonus is granted by the fact that the refunding is paid with the marginal abatement costs, which comply with the grant/t CO_2 only for the last unit. Therefore, the British system, which can claim for a special regulation at present, ought to get into serious difficulties: The EU provisions on state aid actually forbid such benefits for companies.

Denmark confines its system to the generation of power. As a result of this only a few companies with similar abatement costs take part in the trade. Thus, an extreme low trading volume is pre-determined. Accordingly, the hitherto experiences are very poor. The traded quantities are so low that one can hardly speak of a workable market.

The Netherlands strive for a **benchmarking** system. Here the national accepted absolute restriction **(cap)** is not transferred to the company level but only specific caps. In this way an incentive should be given to companies to produce with the most CO_2 efficient technology. According to this **benchmark** technology, the allocation takes place free of charge. Only over and above that definition, certificates must be purchased. However, this system is in the planning stage only. The Dutch government makes several great efforts to realise JI and CDM projects. These are planned and financed on the government's own or by the World Bank in order to fulfil the Kyoto commitments.

3.3 The EU Proposal from October 23rd 2001

3.3.1 *The suggested regulations and their explanations*

The EU Commission published a proposal for a directive establishing a scheme for greenhouse gas emission allowance trading on October 23rd 2001. This presented proposal was preceded by a Green Paper from March 2000, which had provided other regulations for important articles. On the one Hand, modifications of the proposal, which were carried out in summer 2001, indicate the willingness of the commission to discuss. On the other hand, these modifica-

---

[30] These correspond with the area under the marginal costs of reduction.

tions are a sign of considerable methodical problems that are connected with the change to a new emission trading system at the "installation-" or company-level.

The legally binding introduction of a trade system for CO_2 certificates, which is restricted to certain installations and connected with a severe sanction mechanism, is described in the present proposal for a directive. A trade with certificates for CO_2 emissions for certain installations of the producing sectors (basically, the energy conversion sector and the energy-intensive sectors are included) should take place in two periods 2005-2007 and 2008-2012.

The claim is that reductions of greenhouse gas emissions are promoted in a cost-effective manner (article 1). According to this, every installation (power station, cement furnace, ceramics furnace, ...) that is listed in Annex 1 and possesses a certain capacity must hold permits at its disposal from January 1st 2005 onwards, otherwise the activity is prohibited (article 4). As the proposal applies only to CO_2 for the moment, this means a change of all hitherto given operating licenses for fossil fired installations with a relevant capacity.

According to article 9 of the proposal, every Member State must develop a national allocation plan to show the intended steps and to fulfil the national climate commitments. Installation related emissions serve as a basis for the basic equipment. De facto absolute national commitments predetermined in the Kyoto Protocol are transferred to the company level with regard to the installation level. Annex III (1) refers to this. Pre-determined criteria according to Annex III (2)-(7) are of a particular problematic nature as they are contradictory, impede an efficient trading system and provoke numerous lawsuits. For example Annex III (5) says that no installations shall be allocated more allowances than it is likely to need. It is no longer understandable how supply shall come about except by closure and relocation of installations.

The national allocation plans must be presented to the commission for acceptance. In face of the partly contradictory requirements of the basic equipment and the reduction rates that are pre-determined according to the national allocation of commitments in the burden sharing agreement, a potential of conflict is in this point. This fact must have a negative effect on the long-term planning reliability of the energy and energy-intensive companies.

In article 10 it is determined that Member States shall allocate allowances free of charge for the first period 2005 – 2007 according to a grandfathering system. For the second period 2008 – 2012, which corresponds with the Kyoto commitment period, the Commission shall specify a method of allocation. The hitherto discussion in the European Parliament shows that there is at least partly a preference for auctions.

In article 11 it is explained that the national allocation plans must decide upon the limitation for every single installation for the three year period beginning with January 1st 2005 and for the five year period beginning with January 1st 2008. In order to avoid distortions of competition, the allocation plans shall take into account the need to provide access to allowances for new entrants. It is not clear how strategies of expanding companies to acquire allowances by outsourcing or founding of new companies shall be prevented.

According to article 12, Member States shall ensure that allowances can be transferred between persons within the Community without restrictions. However, it is not clear whether this regulation is valid in case of closure or partial closure of an installation.

Article 13 defines that allowances shall be valid for emissions during the first and the second period. Accordingly, it is clear that the special trade system shall also apply to following periods.

According to article 14, the necessary guidelines for monitoring and reporting of emissions are fulfilled. In article 16 penalties are determined assuming that an operator does not surrender sufficient allowances to cover its emissions. The excess emissions penalty shall be either EUR 100 or twice the average market price. During the period 2005-2007, Member States shall apply a lower excess emissions penalty of EUR 50 or twice the average market price. The names of the operators who are in breach of national provisions shall be published.

According to article 17, decisions relating to the allocation of allowances shall be made available to the public.

The articles 18-21 arrange the setting up of appropriate authorities, the establishment of registries and the designation of a Central Administrator in the EU Commission for monitoring the trade system.

An integration of the Kyoto mechanisms was only denoted as "desirable" within the compromise proposal of the Danish presidency from August 28th 2002 (see chapter 7.2). In this way the EU trade system is isolated from the Annex B-wide instrumentation. Consequently, the EU-internal operators of the energy-intensive sectors do not have the possibility to fulfil their indigenous commitments by means of Annex B-wide emissions trading as determined in the Kyoto and Marrakesh agreements or by the realisation of JI or CDM projects.[31] The temporal synchronisation with the realisation of the Kyoto mechanisms is not guaranteed with the determination of a period I from 2005 – 2007, within which sectors covered by the directive are forced to participate in the EU system. In this way, the possibility arises that in the period 2005 – 2007 within the EU high prices apply to parts, that are clearly lower after the subsequent integration of some of the Kyoto mechanisms from the year 2008 on. Either there are high incentives for migration of industries or attentism is induced in the national allocation plans by waiting for the next period II. In any case an outcome of this must be an inefficient solution .

According to the proposal, operators can search for a cost-effective solution within the EU only. The EU wants to reach its Kyoto commitments in a less expensive way by means of the EU "emissions trading" (ET). Whereas, it is quite unclear to which reference system this profitability applies. If an isolated national reduction forms the basis of this reference system so that every single EU country reaches its targets with the aid of indigenous economic instruments and uniform reduction commitments for every partial sector only, an irrelevant benchmark was chosen: On the one hand, practised climate policy in the individual countries is not that "silly". On the other hand, just those flexibilities that make such a policy unnecessary come into force with the implementation of the Kyoto mechanisms. But these are not implemented from the outset.

The term "emissions trading" for a system proposed by the EU is unclear, because a different kind of emissions trading with different participants and different rules has already been arranged in the Kyoto Protocol. It is obvious that just this Kyoto emissions trading shall not be combined with the EU trade within narrow subset of the industry. Therefore, the allowances acquired in the EU emissions trading do not have the same value like allowances acquired in the Kyoto emissions trading. However, one must ask if this confusion of ideas was induced deliberately in order to attack opponents of the special EU concept with the global suspicion of arguing against the Kyoto Protocol. At least it is dubious to use the same term for two distinct matters.

---

[31] The compromise proposal of the Danish presidency from August 28th 2002, which is analysed in more detail in chapter 7.2, denotes project-based measures only as "desirable" and therefore, does not integrate them as legally binding and competitive neutral.

3.3.2 Inner contradictions

The simple economic logic behind an ET system is: If relative differences exist between involved units, efficiency can be increased for all units compared to uniform standards by creating exchange possibilities that provide advantages for at least two participants. Consequently, trade accomplishes gains in efficiency. Therefore, economists look upon it favourably.

a) The estimation of trade advantages

At the quantification of trade advantages two aspects are in equal measure important:

- **A uniform standard, binding upon all parties,** is used as a reference base. Actually, this would be an absurd political approach. In any case, the hitherto climate protection policy paid attention to the different cost situations and technical possibilities as well as to the duration of capital stock reorganisation. Therefore, it was and is not that inefficient as it is imputed in the reference base. Theoretically, it would even be possible for an "omniscient" planner to estimate the efficient contributions approximately exact. This special case does not appear quite realistic. It only serves as a clarification of the underlying arguments: The exaggerated assigning of supposed efficiency gains is provided by an absurd reference base. However, restrictions have to be taken into consideration as the differentiation of the legal framework itself can in turn cause administration and regulation costs.

- Numerous dischargers with **distinct different marginal abatement costs** should be admitted to the "club" of the ET system. The more similar the club members are in this respect the lesser are the potential efficiency gains by trade. In the Heckscher-Ohlin-Model[32] of foreign trade especially the **differences** between economies found the gains of trade. One can start out from the fact that in the case of the Kyoto Emissions Trading, clear differences of marginal abatement costs exist, whereas in the case of the EU proposal, the trade system is focused on a group of participants with very similar technologies in use and, therefore, low advantages from trade. This is valid as long as JI and CDM are not implemented as abatement measures in the trade system, too.

---

[32] See, for example, Dieckheuer (2001), chapter 3 or Ströbele/Wacker (2000), chapter 3.

With regard to a concept that is not used in reality, apparent high efficiency gains by trading with GHG certificates can be calculated with the aid of certain models. This is exactly the point to which the EU proposal refers. On the contrary, a valuation how much additional CO_2 abatement is reachable in an ET system under compliance with leakage restrictions and towards a reference base that takes the actual instrumentation of the already working climate protection policy as a yardstick.

Beside these classic explanations for a certificate system, modern environmental economics professional discussions point more and more to the fact that every instrument (in case of the same aspired environmental effectiveness) shows also **specific costs of its institutional design**. There are, beside measure and monitoring costs, costs for the realisation of a trade system, costs for losses in the form of decreases in production or reduced growth (because of different leakage effects), which can be attributed to the instruments itself.

Because of the strong bond of the energy content to the carbon content, up to now one can do without direct measures of emissions and, instead, might register the use of fossil fuels in combustion processes only. Thus, also the EU Commission refers to it: "Emissions of carbon dioxide are widely recognised as capable of regenerating good quality monitoring data on a consistent basis."[33] This hint in the explanatory memorandum indicates that the Commission has not understood the change of the regime with the transition to an ET system in all its consequences. As long as companies do not see a serious disadvantage in the co-operation with statistic authorities and, on the contrary, have to do a proper accouting out of self-interest because of the costs of the obtained inputs, a problem of measuring will not come into existence. But if the by-product CO_2 is traded more expensive than the input factors and if the specific emission factors within groups of fuels differ by some percentages because of distinct chemical compositions (lignite and hard coal, mineral oil and natural gas are de facto no perfect homogenous fuels), what after all can mean some million € difference regarding costs or financial penalties respectively, this optimistic point of view in the explanatory memorandum is no longer tenable.[34] In view of the estimated magnitudes of certificate prices and financial penalties respectively, a pure CO_2 certificate system needs relative extensive monitoring and enforcement methods in order to be enforcable by law. There-

---

[33] Explanatory memorandum of the proposal, section 10, paragraph 1.
[34] Thus, the emission factors of the Research Centre Jülich differ from those of the IPPC with regard to important fuels. See Birnbaum et al. (1991).

fore, exact emission upper limits for individual companies are ruled out because of considerable bureaucracy costs.

Thus, one can see that the efficiency of a global trade system is considerably restricted by the proposal of the EU Commission. Through a restriction to the energy industry and energy-intensive sectors of the industry prices will arrive that are clearly higher than those of a system that includes at least basically global cost-effective reduction potentials under recourse to the other flexible instruments of the Kyoto Protocol.

b) Advantages and disadvantages of diverse emissions trading systems

However, there is not one single emissions trading system, but there is an extensive range of conceivable solutions each with specific problems. First of all, one has to determine which entities should participate in the trade of GHG emission certificates: According to the logic of the Kyoto Protocol and the EU burden sharing agreement, **countries** in the first place should trade with GHG emissions as countries have agreed upon national reduction commitments. In particular, the flexible Kyoto mechanisms are provided for the involvement of states (state emission trading, state foreign aid for energy projects, ...).

According to the EU proposal, only specific firms as operators of listed installations should be actors in a trade system. Against the textbook conception, not all companies or dischargers of GHGs are involved, but only a subset of these. According to the EU proposal, restricted rules apply to companies that are involved in the EU trade system with regard to the use of the Kyoto mechanisms, certain severe sanctions and inflexible upper limits for each individual discharger within this subset. As there will be in any case an Annex B-wide trade allowed in accordance with the Kyoto mechanisms and the other flexible mechanisms at least for other dischargers, a juxtaposition of different climate policy regimes arises. By establishing the national allocation plans, the national planer would only by chance succeed in breaking down the subsets along with the appropriate reduction requirements so that a conceivable efficient solution arises. The magnitude of CO_2 prices in a range of 20 to 30 €/t CO_2 according to the calculations of the EU and 5 US $/t CO_2 for current CDM projects of the World Bank shows already in the run-up massive inefficiencies. In addition, 15 national planer must co-ordinate allocation plans with requirements like "competition neutrality" or "review of government aid" so that these requirements are compatible with the allocation plans. Basically, this is a hopeless target that must lead to never ending quarrels and creates the opposite of middle- and long-term planning reliability that are essential for the energy sector and energy-intensive industries.

As the EU has the obligation to check the allocation of allowances (free of charge) to companies under the aspect of government aid – otherwise, every Member State could provide its own energy-intensive companies a competitive advantage – high bureaucratic costs would arise by realising the proposal.[35]

The most important reason of the EU to consider a subset of dischargers only is based on the fact that otherwise the monitoring and implementation costs would become extremely high. Therefore, a juxtaposition of two different active sub-systems is created which should achieve the national pre-determined commitments of the burden sharing agreement. Beside the trade system different standards apply to measuring and monitoring – otherwise the limitation does not make sense. This leads to an absurd incentive for encouraging companies or parts of companies to search for advantages in both systems. As every firm, for example, buys fuel oil for normal heat processes or energy sources for certain processes and installations, which are not covered in the EU proposal, the use of energy sources must be kept under surveillance. Or else the EU must transfer the same monitoring and implementation standards to other fields. In this case the explanation for the division in two sub-systems is invalid.

c) Allocation methods for the basic equipment

At an emissions trading on company-level an allocation method for the basic equipment of emission certificates is necessary. On the whole, there are four methods with different effects:

(1) The sale of emission certificates by the state at fixed prices with and without allowances. This corresponds in its effects with a carbon tax with and without allowances. Therefore, this approach is not discussed in literature on certificate trading schemes although it is systematically conceivable.

(2) Auctioning of certificates that is carried out in regular intervals for each subsequent period (e. g. five years) leads to temporary restricted emission rights.

---

[35] These special incentives result from different treatments of the ET sector and the non ET sector: Merely the missing sanctions for the Member States for the non-achievement of national reduction

(3) Allocation of certificates free of charge according to actual emissions of a base year, i. e. so-called grandfathering connected with a depreciation for the periods to come.

(4) Allocation of allowances free of charge according to technical efficiency criteria, i. e. grandfathering according to reference technologies. On the one hand, energy efficiency can serve as a standard, on the other hand, specific CO_2 emissions can be used.

In case of two-digit certificate prices (€/t CO_2) that are mentioned in the EU proposal, auctions would de facto lead to a partial expropriation of installations using lignite or hard coal. Economic as well as legal problems, which partially hold the rank of constitutional principles like, for example, the protection of property or equal treatment, are combined with this fact. Hitherto unrestricted valid permissions would effectively be abolished or strongly restricted. Auctions raise considerable legal problems insofar as they make the legitimacy of such allocation methods uncertain. However, on October 10th 2002 the EU Parliament argued for the fact that already in the first period (2005-2007) 15 % of the emission certificates shall be allocated in return for payment, for example, by means of an auction and that the remainder shall be allocated free of charge. The EU Commission has already argued for auctions in its Green Book.

In case of a cap-and-trade-system an appropriate **grandfathering method** has to be found, which is accompanied by several problems (again these problems are hardly discussed in textbooks). As, according to the EU proposal, an absolute upper limit for CO_2 emissions should apply for each individual installation, there must be some kind of a "classic" grandfathering. But the determination of a base year and a reduction path for every individual company evokes considerable problems:

If the certificate prices (mentioned in the EU proposal) are sufficiently high, the specific determination of emission limits free of charge causes a consolidation of the market share in the base year. De facto the EU would have reproduced the mechanism with which the so called "seven sisters" had internally stabilised their cartel on the world oil market in the fifties and sixties: At that time growth by innovation or a very good marketing were sanctioned with financial penalties so that there were no incentives to deviate from the given

---

commitments and very serious financial penalties for companies in the ET sector lead to two distinct systems.

market shares. From the point of view of competition policy, this is hardly acceptable. A system that seems to be only efficient from an environmental policy point of view creates **inefficiencies and competitive barriers on the product market**.

Companies that enter the market (newcomer) should not be discriminated according to the logic of market-economy and they could at least lay claim to certificates free of charge as allocated to their established competitors. For these reasons, every hitherto existing company can claim for new certificates through outsourcing and founding of new subsidiaries. Companies that had already planned an installation in the base year or shortly before, which is well utilised and energetically optimised today, are the losers in this game. The certificates are allocated according to the utilisation of the installation and as a result of this, companies get less allowances if the installation was not optimally utilised in the base year.

Foreign competitors in growing markets enjoy a bestowed competitive advantage. However, this does not help climate protection. In shrinking or stagnating markets indigenous companies receive closure rewards. These are not the expected signals according to a market-economy.

The term "installation" is not sufficiently defined in the EU proposal: If installations are seen in the view of comparable outputs and neutral competitive basic equipments, all electricity generating installations should get the same amount of CO_2 certificates on the basis of kWh: Lignite, natural gas and oil power stations in equal measure. As this would hardly be acceptable, differences in "installations" should be allowed that would lead to a separate allocation of allowances. On the other hand, this can lead to different treatments of hard coal power stations because of a magnitude of criteria like, for example, a different furnace technology: In comparison with conventional grate firing, fluidised bed power plants should be treated as different installations. In the annex of the EU Proposal are no technical criteria how this problem has to be handled. Factually, this problem can only be solved by an instantaneous value of a base year.

Another procedure would be to issue the basic equipment according to uniform technical standards. Accordingly, for example, every lignite power plant, every natural gas power station and every cement kiln would receive a basic equipment according to a "good" or best state of the art of a base year. This can justly be regarded as "competitively neutral". As the national reduction commitments according to the burden sharing agreement would then develop differently, divergent specifications according to the technical state of the art will arise for future times. It is not comprehensible how these contradictions can be clarified so that a system of absolute caps-and-trade can be designed.

On the other hand, a "competitively neutral" implementation is demanded, hard coal power stations of the same technology must receive the same number of certificates per generated unit of current and, namely, regardless of whether the installation is in Germany, Greek or Ireland. Then a system would be created that is based on the allocation of certificates free of charge according to technical efficiency criteria. The total amount of the national pre-determined emissions must then be fulfilled through the contributions of the other sectors and through the use of the Kyoto mechanisms. However, such a system is not compatible with the EU proposal. The requirement for a "competitively neutral" allocation of certificates listed in Annex 3 can not be achieved under the given framework of the proposal.

3.3.3 Decision procedure

The Commission based the proposal on the environmental competencies of the EU according to article 175, para. 1 ECT implying the co-decision procedure according to article 251 ECT. Correspondingly, the Commission presented the proposal to the European Parliament, the Council (Environment) and to the Economic and Social Committee on October 23rd 2001. Furthermore, the Commission asks the Committee of Regions to make a statement about the proposal. The Environmental Committee is leadingly responsible in the European Parliament. This Committee will present its statement with a modification proposal to the Parliament in autumn 2002. After working groups had paid attention to the proposal, the Permanent Representatives Committee drew up a preliminary report in the Council of Ministers on February 18th 2002.

In the course of the process the Council of Ministers makes a decision about the proposed changes of the European Parliament. If the parliament does not carry out changes, the Council of Ministers can decide on the proposal with a qualified majority. Then the draft directive is enacted. If the Council of Ministers requires further changes or does not accept changes of the European Parliament, it will draw up a joint position which will be transferred to the European Parliament. Now the European Parliament can accept the changes of the Council of Ministers with an absolute majority in the second reading, then the act of law is enacted, or reject it with an absolute majority with which the directive would not be enacted and the proceeding would be finished. However, if the European Parliament decides on the proposal with an absolute majority, these changes will be forwarded to the Commission, which will make a statement.

If the Commission approves the changes of the European Parliament, the Council of Ministers can enact the directive with a qualified majority. In case of a rejection of the changes of the European Parliament by the Commission, the Council of Ministers can only come to an unanimous decision on the changes of the parliament. If the council does not accept all changes, a Conciliation Committee will be called. In case of a discrepancy between the Council and the Parliament, the guideline will not be enacted. In case of an agreement in the Conciliation Committee, the compromise must be accepted in the Council with a qualified majority and in the Parliament with an absolute majority of the votes.

Until now it is not clear whether article 175, para. 1 ECT is the appropriate legal basis for the proposal. As such a proposal influences strongly the Member States' choices of energy sources and the general structure of energy supply, which result from an estimated certificate price of 33 €/t CO_2 determined by the Commission itself, article 175, para. 2 ECT would also be an appropriate legal basis. In this case by means of a consultation procedure, the Council would come to an unanimous decision on the draft directive after having listened to the Parliament. Beyond advising comments the Parliament does not have any possibility to reject or accept the draft directive. It's only up to the Council to decide.

A further change of the decision procedure can be necessary if the introduction of taxes, that are not that clearly defined in the EU, is implemented in the EU trade system. This would be the case, if the basic equipment is issued by auctions.

It is not possible to discuss in detail to what extent article 175, para. 1 or para. 2 EGV are the right legal basis of the proposal and how a possible change of the legal basis can be carried out.

3.4 EU Directive and energy policy

Factually, a reduction of CO_2 emissions can be achieved in four major ways:

1) Increase of energy efficiency on all stages of energy use and conversion, i. e. **general "energy saving"** reduces energy consumption compared to a "business as usual" path. Innovation of procedures as well as substitution processes between energy inputs and capital expenditures (better heat in-

sulation of buildings, elaborate motor technology, better control of electric motors, heat recovery plants in the industry, ...) are necessary to reach this target.

2) Relocation of energy-intensive and, above all, of CO_2-intensive processes to **foreign countries (leakage)** that is not that restricted by such a strict system. Practically, this does not help the world climate at all because of the global character of the greenhouse gases. Furthermore, this can even lead to higher greenhouse gas emissions because of worse technologies abroad and higher shipment volumes (cement from Taiwan to Central Europe?) but it improves the European emission balance.

3) Change of the **energy mix** reduces the CO_2 intensity of production: In the generation of electricity this would promote the use of nuclear fission instead of coal generated electricity. In heat processes this would require an increased use of natural gas instead of mineral oil or coal.

4) Change of **consumption patterns** can lower the energy consumption. This concerns, for example, the hitherto use of energy-intensive systems like transportation systems (partly transition of traffic from road to rail or inland waterways, style of driving, substitution of German air traffic with other high-speed transportation systems, ...) or the supply of "heated rooms" where still a lack of knowledge exists (wrong ventilation of rooms, high room temperatures, covering of radiators, ...). In this context only price signals and better information can help.

Whereas the first statement was and will be an important aspect of the hitherto strategy, the second statement, however, can hardly be envisaged. Thus, the question arises in how far the third statement is of importance. In the face of the importance of economic efficiency in the years to come, this means, in case of the realisation of the EU proposal, factually a strategy that increases the share of nuclear energy and natural gas in the primary energy production and that reduces, above all, the share of lignite and hard coal. Thereby, however, the energy mix of the economies will only depend on few energy sources, which can give rise to **conflicts with security of supply objectives:** Nuclear energy can factually make only contributions to the base load supply of electricity, which would therefore depend on nuclear energy only (beside hydroelectric power plants); only few countries produce natural gas for Europe whose market power would increase considerably if, for instance, the Member States of the EU increase the market share of natural gas considerably. In Germany - after its phasing out of nuclear energy - this strategy must cause a clear increase of imports of nuclear energy and natural gas. But changing the energy mix of the Member States requires a different decision procedure according to article 175, para. 2 ECT.

With regard to the primary and industrial goods industry as well as the conversion sector statement 4 should not play an important role. However, it is of importance to the sectors of transportation, households, and small consumers.

4 Assessment of effects of the EU draft directive

4.1 General aspects

The economic effects of an emission trading system are decisively affected by the determined allowance prices which are calculated by the use of formal methods. These prices, for their part, are depended on the CO_2 abatement costs and the pre-determined targets. Because of progressively increasing marginal abatement costs, total costs and therefore allowance prices are increasing with the extent of the underlying commitments.

	Alternative I EU draft directive PRIMES u.a.	Alternative II EU draft directive *code* TM	Alternative III Kyoto/Marrakesh *code* TM
Implemented in the model		• *Leakage* effects • Natural gas price increase	• *Leakage* effects • Natural gas price increase • Rules of Marrakesh • JI, CDM, Kyoto-compliant emissions trading • Strategic behaviour of Russia
Not implemented in the model	• *Leakage* effects • Natural gas price increase • Rules of Marrakesh • JI, CDM, state emissions trading	• Rules of Marrakesh • JI, CDM, state emissions trading	

Scheme 2: Survey of modelled alternatives

The modelling includes two phases: First of all, a prediction for CO_2 abatement effects and CO_2 prices will be given on the basis of the Applied General Equilibrium model *code* TM, which covers the whole world as a climate economic model. This step includes an isolated EU trade system as it is provided in the present proposal. In the sectoral analysis of chapter 6, which can be done more disaggregated and can insofar reproduce sectoral effects more precisely, CO_2 prices that were endogenously calculated in chapter 5 are given. At first, the sectoral analysis takes the CO_2 prices mentioned in the explanatory memorandum of the proposal of the EU Commission as a basis. Consequently, the model calculations for alternative 1 (isolated EU system according to EU estimations) act on the assumption of 30 €/t CO_2. However, the following simulations show that important retroactive effects on the sectoral production structure are not included in the CO_2 price determined by the EU Commission. Thus, at a price of 33 €/t CO_2 Germany is a net supplier of certificates which can not be verified in model calculations. On the other hand, important effects like the relocation of energy-intensive industries or a large increase of natural gas prices are not explicitly implemented in the EU model. However, the EU calculations are taken as a basis for first model simulations in order to reveal contradictions of the EU explanations.

In the calculation of price effects diverse secondary effects have to be taken into consideration, which are possibly inadequately modelled in the economic approaches. If, for example, the model is based on a CO_2 certificate price of 30 €/t and if only limited fuel substitution possibilities of the European industry are implemented, two major effects are not reckoned: On the one hand, the use of hard coal would more than double in price, so that serious leakage effects must arise. On the other hand, an intense fuel substitution from coal to natural gas would be set off. Hence the producer price of natural gas would rise. These effects would clearly reduce the actual CO_2 prices – however, at the expense of an increase of CO_2 emissions outside the EU. The production of energy-intensive goods would then take place in the USA or in Taiwan, the generation of electricity would be shifted to Eastern European states. This would not help climate protection but there would be an apparent statistical success for the EU. According to our model calculations, **CO_2 prices** around 15 €/t CO_2 arise even in an isolated trade system of the EU because of leakage effects and an increase in the price of natural gas (alternative II: EU system without implementation of Kyoto mechanisms). It has to be pointed out that models which do not explicitly implement leakage effects have to calculate higher prices for CO_2 allowances. This is because companies with high abatement costs do not have the possibility to relocate their installations. However, if this option is taken into account, the calculated price must fall, but this is based on the relocation of companies and must therefore be paid with the loss of indigenous jobs.

Furthermore, it is crucial if an isolated trade system is taken into account or if the flexible instruments of the Kyoto Protocol will be integrated in a trade system. If these cost-effective abatement options are implemented, a broader choice of action is possible for climate protection and as a result low CO_2 prices will show up.[36] In order to be able to calculate this case which has already been predetermined on the part of international law, we take CO_2 prices of 5 €/t CO_2 (alternative III) as a basis because of the results of our model simulations. It is important to point out that this alternative is neither defined mandatory by the EU draft directive nor by the Danish compromise proposal from August 28th 2002.

The market reforms of network supplied energy sources (electricity and natural gas) since 1999/2000 is another important aspect. The development of liberalised markets changes the price formation in power industries. Prices are no longer determined by the average costs of power companies that were previously organised in local monopolies. Now supply competition (or in the British system from the year 2000: the bidder competition for the energy supply that is demanded from the pool) determines the relevant electricity prices at half-hour or quarter of an hour intervals. According to the experiences of the British system, this electricity price is clearly determined by the marginal coal-fired power stations.

Hydroelectric and nuclear power stations operate cost-effectively subject to their technical availability. Gas and steam power stations of independent producers (at least in Great Britain) are also cost-effective as base load electricity suppliers. In Germany existing lignite power stations would also act as base load suppliers at least up to the amount of allowances that was allocated free of charge. In Germany, renewable energy sources must be fed in at fixed prices according to the Renewable Energy Sources Act: consequently, their supply is exogenous.

---

[36] Thus, for example, a study carried out by Böhringer/Rutherford (2000) shows that, on average, the marginal abatement costs are around 170 $/t CO_2 in Western Europe and, on the contrary, only around 11 $/t CO_2 in Eastern Europe.

The necessary balancing of supply and demand is up to average load power stations that work on the basis of hard coal and eventually also natural gas in future so that price determination at the production stage is defined by this. Consequently, if the marginal unit of hard coal is charged a CO_2 price of around 30 €/t CO_2, this leads directly to a serious increase in electricity prices at the production stage. This effect would not only have consequences for the energy-intensive industries with high specific energy consumption patterns but also for other parts of the economy and households. Winners would be the operators of nuclear energy power stations in Europe as well as natural gas producers by windfall profits.

With regard to the technical realisation of the model the question arises how the calibration of an Applied General Equilibrium model with a base year in the middle of the nineties can take account of this changed institutional structure: the model already contains the institutional and legal framework, the responses and patterns of behaviour that are reflected in the data of the nineties, which are no longer relevant today and, above all, will not be relevant in future. Models of environmental economists who are not familiar with the new conditions of the liberalised energy markets are on shaky ground. Political changes like, for example, a complete reform of the markets or a phasing out of nuclear energy must be integrated into the model separately.

4.2 EU assessment of CO_2 prices

The impact assessment form in the annex of the EU proposal for establishing a scheme for greenhouse gas (GHG) emission allowance trading is based on the EU financed models PRIMES (in connection with POLES) and GENESIS.[37] Because of the simulations of these different model types, the spectrum of the expected allowance prices reaches from 20 € to 33 €.[38] In comparison with own model simulations and with results of other research groups, this cost estimation for a scenario presented in the EU proposal can be judged as quite realistic. Of course modifications that were carried out after the Marrakesh conference concerning the CO_2 targets have not been taken into consideration yet. If the EU proposal would be implemented as proposed, a first estimation of the consequences can be carried out on this basis. Hence, the effects of an

---

[37] Descriptions of these models are in Blok/de Jager/Hendriks (2001) and Blok et al (2001). A detailed description of the scenario for PRIMES is in Capros/Mantzos (2000).

[38] See EU Commission (2001), p. 50, fn 2 & 3.

implemented system according to the EU proposal with its estimated prices should be determined first of all.

In addition, it is of main importance to the total sectoral financial burden whether emission allowances have to be bought at the market for the entire amount of emissions or only for the amount of emissions required to reach the reduction commitments. In the first case, CO_2 emissions are traded altogether and in the second case, only the CO_2 abatement is traded.

As already mentioned, the EU proposal leaves open at least for the second period to which extent the sectors involved in the trade must acquire emissions allowances. In the Green Book an auctioning system is preferred that would force market participants to proof that all emissions are covered by appropriate allowances. Furthermore, there are no information about the future price path of emissions allowances so that there are no information of the amount of transactions and prices. The sectors determined in the EU proposal, for example, are obliged to acquire allowances for the total emissions of 338,5 Mill. t CO_2 at a price of 33 € with additional costs of 11,17 bill. €. These additional costs would be reduced to a tenth if the commitment only refers to the difference between reduction target (about 300 mill. t CO_2) and actual emission volume (status 1999: 338,5 mill. t). In the face of these uncertainties, the estimated effects must be seen with considerable reservations.

4.3 Critical appraisal of the EU assessments

The use of the model PRIMES is problematical outside this tightly defined scenario, as in the case of modifications of single aspects like the implementation of flexible instruments.

Despite probable results according to the restricted scenario of the EU proposal, some methodical aspects have to be pointed out that represent a clear source of errors and suggest care with regard to the applicability of model simulations to direct political consultancy. In connection with climate protection restrictions, the following aspects are graded as deficiencies: Partial analysis, sectoral aggregation and dynamics.

A general defect of the used model is caused by the inadequate implementation of costs that arise due to structural and regional unemployment and thus, resulting structural-crises. Such have to be taken into consideration in case of a serious decrease of lignite-based electricity generation in the mining area on the left bank of the Rhine or in East Germany or in case of a drastic decrease of the use of hard coal in power stations in West Germany. At least in these cases additional labour market policy options have to be taken into account.

➢ Partial analysis and trade relations: producer prices

Only the territories of the EU are implemented in PRIMES whereas the rest of the world is exogenous determined (parameterised). As a result of this, *feedback* effects, which are of great importance to climate protection, are not taken into consideration.

Producer prices will change if countries outside the EU are subjected to climate protection as well. Within the Annex B countries the demand for coal will decrease substantially and the demand for mineral oil products will decrease slightly. Thus, supply prices for hard coal on the world market fall. The oil price is only slightly affected, whereas the producer price for natural gas increase. But the implementation of the Kyoto mechanisms will change the demand of energy.

In this case the instruments Emissions Trading (ET) and Joint Implementation (JI) have to be taken into consideration. Thus, for example, Russia will abate CO_2 emissions because of its integration in an ET system, which will have a significant influence on the use of fossil energy sources there. As Russia is a main supplier of the European natural gas imports, the Russian increase in consumption must have an impact on the EU prices on a medium-term basis. This is of vital importance to the substitution processes between fossil energy sources and therefore to the price of CO_2.

The implementation of JI projects within Annex B-countries or of CDM projects outside Annex B-countries has clear impacts on the consumption of fossil energy sources in the project countries, too. Above all, these effects can clearly be seen in homogenous goods, such as crude oil, hard coal and LNG, and partly in pipeline gas, which can be classified as a continental homogenous good. Because of the price effects in different markets important feedback effects arise for CO_2 prices. Even a combination of POLES, a world trade model developed for the EU in Grenoble, and PRIMES can not solve this problem satisfactory as the decision to use fossil energy sources has to taken be simultaneously in all regions in the case of changed CO_2 restrictions. Consequently, the regional limitation of PRIMES make the modelling of feedback effects impossible, which are of great importance for the determination of CO_2 prices.

➤ Partial analysis and trade relations: Leakage effects

In the model simulations of the EU relocation of industries to countries without climate protection restrictions can only be taken into account by means of exogenous factors. Thus, the explanatory power of an Applied General Equilibrium model is considerably restricted. According to the above-mentioned comments, the minimisation of leakage effects, however, represents an important condition for a successful climate protection policy because of reasons to preserve national growth potentials and employment. However, exactly this point is not modelled explicitly.

➤ Partial analysis and flexible instruments

The Marrakesh Accords provide the framework for trade of emission allowances between the EU and the Umbrella Group[39], which was reduced by the withdrawal of the USA. Furthermore, the opportunity exists to credit sinks of CO_2 and to use JI or CDM to comply with the commitments. On the contrary, the EU proposal disregards explicitly the EU-wide application of these flexible instruments in the first period in contrast to the second period (2008-2012). In PRIMES none of theses flexible instruments is implemented. The range of results (20 to 33 €) is based on the fact that the CO_2 commitments of the EU of about 8 % must be achieved with indigenous measures only. A multitude of model simulations shows that the allowance prices will fall clearly if ET and JI are applicable beyond EU borders. PRIMES is not able to reproduce the global implementation of the flexible instruments that was determined in the Marrakesh Accords.

➤ Sectoral aggregation

The particularly detailed aggregation level of the energy sector, seems to be quite advantageous in modelling real processes. But in connection with a top down approach as implemented in PRIMES a problem arises: The consideration of average values at the top-level aggregation leads to numerous inaccuracies or deviations respectively. If a lower aggregation level is calibrated on this basis, these small deviations can lead to a complete activation or deactivation

---

[39] The Umbrella Group consists of Russia, Canada, Japan, Australia, and New Zealand. Kazakhstan filed a petition for participation in an emission trading system, but neither conditions nor decisions were made at the COP 7.

of individual technologies in case of a scenario analysis.[40] But in reality a mix of technologies exists, which can not be activated or deactivated ad hoc. Instead these technologies are often kept active for various reasons. Elasticities of substitution are the basis of substitution processes in General Equilibrium models. But in this respect in the case of PRIMES any documentation is missing. In principle the implementation of disaggregated sectors in a top down approach is always connected to the above presented problems.

➢ Dynamics and the Kyoto commitment

GHG emissions have to be reduced by 8 % according to the Kyoto commitment. The year 1990 serves as the reference year and the target has to be fulfilled in a period 2008 to 2012 on average. The type of dynamic implementation is not revealed in PRIMES. In the year 2010 the target of -8 % is met. If emissions of the year 2008 are still higher than that, they have to be curbed in the year 2012. As a result, a superproportional rise of the CO_2 price is necessary compared to the underlying economic growth. If the Kyoto targets are already met in the year 2008 and emissions are kept constant until 2012, the CO_2 price represents an average of the Kyoto period and increases only because of overall economic dynamics.

➢ Dynamics and the implementation of new technologies

Basically the implementation of strategic behaviour is only possible in optimisation models. Applied General Equilibrium models (AGE models) are not able to implement strategic behaviour. Thus, a so called forward-looking approach is necessary as an approximation. This means, all periods are considered simultaneously to calculate the new equilibrium which is done in a fully dynamic AGE model. On the contrary PRIMES is a recursive dynamic model only, which calculates the optimal behaviour for each period separately. Thus no strategic behaviour can be implemented. There is no documentation on PRIMES concerning this contradiction.

It seems to be quite likely, that the modeller's statement of having implemented strategic behaviour is based on the separation of the electricity sector in three different types of firms and on the assumption of imperfect competition. But if this had been modelled explicitly, there would be an effect on the

---

[40] An explicitly chosen focus of PRIMES is the implementation of the expected technological change. Beside 148 existing technologies for thermal power stations, further 678 expected future technologies are modelled; see Capros et al, p. 3-2.

use of fossil energy resources and therefore on the resulting price of CO_2 emissions.

➤ Dynamics and the use of energy resources

In PRIMES the price paths of fossil energy resources is divided into four periods. In the fourth period, the Hotelling rule becomes active, according to which the royalty – the additional rent resulting from the scarcity of the resources – grows with the individual rate of time preference. This can not be implemented in a recursive dynamic model. It remains unclear in the documentation how this problem was solved. The obvious solution to this problem is an exogenously given price path, but there would be no theoretical consistent explanation of the price level and no empirical validation. But above all, there would be a distinct effect on the use of fossil energy resources and because of that on the price of CO_2 and on producer prices. However, it remains unclear if this aspect was implemented in the model that was used to calculate the results presented in the EU proposal.

➤ The assumption of full employment

The problem of unemployment is neglected in the model used by the EU. But this is far from reality, because the factor "labour" is assumed to be homogenous and almost perfectly mobile. The model does not differentiate between specific abilities (e.g. the driver of a brown coal excavator), regional ties which cause separate costs if abolished, regional negative feedback effects by a fast structural change and the associated regional adjustment costs. If these effects are considered too, the EU statement "no major effects are expected to result from the implementation of the proposal" in the Impact Assessment Form does not appear to be close to reality. Even the hint at additional employees in the trade system itself is not useful: Administration of additional employment in a bureaucratic system can hardly be the goal of the proposal.

5 Model analysis and comparison of scenarios

5.1 Model structure *code* TM

To estimate the economic and ecological effects of the EU proposal the model *code* TM (*Climate policy scenarios in a dynamic general equilibrium trade model*) is used. This full dynamic Applied General Equilibrium (AGE) model incorporates the division of Emissions Trading in two separated systems as defined in the EU-Proposal. The trade model aggregates the world economy into 13 regions and each economy in 13 sectors.[41] According to the climate change problem fossil fuels, energy production, and energy intensive sectors are incorporated in a disaggregated way. The ability of the model to describe global trade flows is important, as by this the leakage effects can be analysed.

5.2 Reference case

In principle, a reference case has to be defined to allow the estimation of economic and ecological effects. This offers the possibility to estimate political options on the same base line. A significant part of the reduction commitment has been done since 1990 and some member states have already implemented different measures to reduce their GHG emissions. These developments have to be integrated in the reference case according to their meaning for the analysis of the model scenarios.

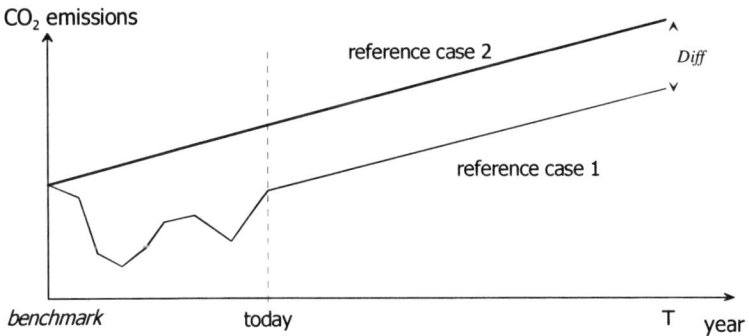

Figure 4: Influence of known data on the reference path

---

[41] Cf. Annex 1. There a model description of *code* TM can be found. For detailed description see Smajgl (2002).

Higher emissions in the reference case produce a higher amount of reductions which are left for the scenario. This incorporates higher reduction costs in the scenario. Figure 4 shows how the reference case is influenced by the incorporation of well known data of the past. The reduction costs of the first reference case of figure 4 are lower than those of the second reference case.

The implementation of relevant data has a significant meaning for the carbon price path. In this study the global Input-output-data with the trade flows are used as the statistical basis. Data for the emission paths are taken from the UNFCCC. This official statistic is connected with forecasts of the OECD about the future demand in fossil fuels for each region to get the dynamic paths. In connection with the specific employment of fossil fuels this leads to the macro economic growth rate of each region, shown in table 2 as an average of the years from 1995 to 2012.

Region	%	Region	%	Region	%
Germany:	2,1	USA:	2,6	Japan:	1,2
UK:	2,2	Canada:	2,8		
Rest of Western Europe:	2,1	Rest of America:	4,4	Non-Annex B-Asien:	5,9
East Europe:	2,2	Australia/New Zealand:	1,7	Middle East/North Africa:	
Former Soviet Union:	1,7			Rest of the World:	

Table 2: Endogenous resulting growth rates of GDP between 1995 and 2012[42]

To be able to analyse the impact of the EU proposal thirteen sectors are defined. Special energy sectors (electricity, crude oil, oil products, natural gas, hard coal, and brown coal) and energy intensive sectors (paper, steel- and metal, and cement, glass and ceramic) are modelled in a disaggregated way.

The full dynamic modelling presents a simultaneous incorporation of all periods and allows the analysis of effects which occur before or after the initiation of carbon restrictions. This corresponds to expectations in investment decisions and is a more consistent modeling technique than in recursive approaches.

In addition, supply of crude oil is modelled in a realistic way by implementing the scarcity of this exhaustible resource. This scarcity leads to a rising price

---

[42] The values of the European regions were taken from estimations of the EU, on which PRIMES is based as well. Otherwise, the results of the model would not be comparable. The growth rate of the other regions stems from estimations of the OECD. For the used grammalogues of the regions, please compare Annex 1.

path with time and induces a substitution within each national energy mix. This substitution process has to be separated from the substitution process induced by the climate change restriction.

In the following, the carbon emissions are described. In Germany the green taxes are implemented. The subsequent reduction in carbon emissions, however, will partly be compensated for by the future growth and the decision to withdraw from the nuclear energy program. Already the latter effect will have an impact of additional emissions of 70 Mt CO_2 in the year 2012 if no additional measures are to be taken. In the end, Germany will be able to reduce the CO_2 emissions until 2008/12 in comparison to 1990 by 3.4%. In the reference case, the UK will have a reduction in their CO_2 emissions of 2.5%. The region „Rest of Western Europe" (REU) will extend the amount of CO_2 emissions by 1.4%. Table 3 shows the changes in the reference case and confronts them with the commitments of Marrakech and the BSA.

reference case	absolute emissions 2008/12 (Mt)	Marrakech commitment (Mt)	relative change from 1990 to 2008/12	Marrakech/BSA commitment relative
Germany	980	801	-3,4%	-21%
UK	603	511	-2,5%	-12,5%
Rest of Western Europe	2101	1846	+1,4%	-0%
Former Soviet Union	2367	3305	-35,1%	+4%
Eastern Europe	943	1004	-3,5%	+5,6%
Japan	1326	1101	+17,9%	-2,1%
Canada	607	471	+30,4%	+1,2%
Australia	375	328	+23,7%	+8,0%

Table 3: Emissions in reference case in comparison with Marrakech/BSA commitments[43]

The emission paths of the other Annex B-countries are significantly higher. For example, Canada's increase in CO_2 emissions between 1990 and 2008/12 lies at 30.4%, that of Japan by 17.5%, and that of Australia by 23.7%. The Former Soviet Union (FSU) boasts significantly positive growth rates, but still shows a huge amount of Hot air in 2008/12. Therefore, the reduction of CO_2 emissions of FSU is 2008/12 35.1% lower than in 1990. Eastern Europe also shows small amounts of Hot air as well as a reduction of CO_2 emissions of 3.5% in the same period.

---

[43] As it was declared in several political statements carbon sinks are not incorporated for the EU.

In the year 2012 there is an average amount of Hot air of about 1.000 Mt CO_2. 94% of this amount originate in the Former Soviet Union. This supply of Hot air would be confronted in a Kyoto-conform Emissions Trading scheme with the following demand: Germany will need an additional reduction of nearly 180 Mt CO_2, UK of 90 Mt CO_2 and the region Rest of Europe nearly 260 Mt CO_2. Canada will have to reduce an additional amount of more than 135 Mt CO_2, Japan of up to 225 Mt and Australia of nearly 50 Mt. These data emerge, if CO_2 is reduced proportionally to the GHG reduction commitment.

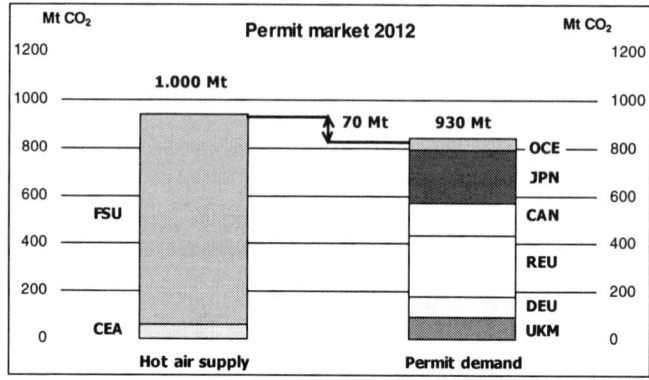

Figure 5: Market for CO_2 emission permits in the year 2012.[44]

Figure 5 confronts the estimated amounts of supply with the demand for 2012. In the average of 2008/12 there is an over supply of about 70 Mt CO_2. At this point three aspects are obvious:

- The price of CO_2 will be low depending on how a certain region reacts to the restriction.

- The withdrawal of the USA leads to an economic advantage for the remaining Annex B-parties. The huge demand of the USA for about 1,135 Mt CO_2 for the year 2012 is kept out of the market of CO_2 emissions permits.

- The owners of Hot air experiences a significant economic loss due to the withdrawal of the USA.

---

[44] The country codes are listed in Annex 1.

It has to be noted that a higher growth rate in Annex B-countries would lead to an increasing demand with a simultaneous decrease of the remaining supply. In such a case the carbon prices would be higher.

In addition, the character of Hot air has to be clarified. The available amount of Hot air decreases due to the growing economy. After a certain period of time this amount will be consumed by the domestic commitment. Therefore the Hot air effect on permit prices has a short-term character and will disappear sometime between 2015 and 2018. From this point on the existing reduction costs will determine the permit prices which means that the prices will rise significantly. Long-term sectoral effects are analysed in chapter 6.

The described reference case is used as a basis for comparing different concepts for an Emissions Trading scheme. In comparison to this reference case, in chapter 6, we will assume the implementation of an Emissions Trading scheme, see chapter 6.1. The reference case used in that part will also be developed in the following scenarios. At first, the German situation has to be examined because the Emissions Trading scheme proposed by the European Commission uses absolute caps. In connection with the base year 1990 and as early actions have to be considered, the discussion about German wall fall profits has to be analysed to define sectoral reduction targets.

5.3 Special case Germany

Absolute caps as they are defined in the EU proposal need a base year to calculate the reduction path until the Kyoto period 2008/12. Within the BSA, Germany accepted a commitment to reduce its national emissions by 21% in comparison to the year 1990. One important reason to accept such a restriction was the decrease of production in East Germany. The question is how this change has to be implemented in the definition of sectoral reduction commitments which are needed in the national allocation plan of the EU proposal. This definition is necessary for the model simulations.

The emission reductions caused by the decrease in production and the former bad CO_2 efficiency have to be divided in two groups to allow a correct integration in the national allocation plan:

a) One part was financed by companies which owned or bought installations and renewed the installations without subsidies. These measures allowed for an increase in CO_2 efficiency and these companies have to get the credit for the absolute emission reductions since 1990.

b) The rest has to be defined as a national amount of emission reductions. This amount directly lowers the national commitment. It has to be divided into two subgroups:

One subgroup was financed by public interventions. As, in this case, tax revenues are used to increase CO_2 efficiency, the absolute reductions have to be attributed to the national emission balance.

The second subgroup is real Hot air which results from companies that today do not longer exist. This amount has to be attributed to national commitment.

As one can see, that the so called *Wall fall profits* were not for free but were realised by a huge amount of tax revenues and private investments. The question is how the two groups can be incorporated into the national allocation plan. In fact, the national value for the base year has to be reduced by the national amounts (group b). The following calculation has to be made according the technical standard. First the national base year's emissions of 1,215 Mt CO_2-equivalent have to be reduced by around 100 Mt CO_2. Additional reductions of about 60 Mt CO_2-equivalents were realised by public investment in the area of other GHG.

The new base year value of 1,053 Mt CO_2-equivalent leads to the adjusted reduction commitment of nearly 9%, which includes the effect of the German Unification. This national reduction path is the base for further calculations to calculate the sectoral commitments. The voluntary commitment of the German industry displays the reduction potential which is technically possible. Thus, an orientation at the fixing of the sectoral reduction path at the high reduction values of the voluntary agreement is legitimate. This approach would lead to an reduction commitment above the national average of 9%, for instance of 12%. This means an under average commitment for the rest sector (households, traffic, ...) of 5,0%. Map 3 clarifies the approach.

(Mt)	Actual value 1990	Reunion effect	Technical standard 1990	Target 2010	Actual value 1999	Remaining reduction
national balance	1215	162	1053	960	982	22
	−21% Reduktionsvorgabe			−9%		
Sectors within the ET system	−12,0%					
Rest of the German economy	−4,8%					

Map 3: Possible Grandfathering approach on base of the technical standard of 1990 and reduction commitments until 2010 in the special case of Germany.

For the level of installations, this approach implies a reduction commitment of 12% in comparison to 1990 and 2008/2012. This will be assumed for the following scenarios. Additionally, the companies in the EU will be equipped with the amount of permits of the base year 1990 via Grandfathering. By fixing a necessary base year in a system with absolute caps, the companies have to be equipped with the amount of permits which is not necessary if early action would be considered. This approach is necessary to allow a consistent implementation of a base year. Additionally, within a different approach it would be necessary to consider variations in the use of existing capacities as well, which would contradict a system of absolute caps.

After the initial allocation the permits will be devaluated by 12% until 2008/12. The rest of the economy is not integrated in the Emissions Trading scheme and therefore does not receive a direct price signal by this system. The government is responsible for this part of the economy and is allowed to use the Kyoto mechanisms, the Annex B-wide ET, JI, and CDM, to fulfill the national commitment.

5.4 Scenario structure

The policy scenarios are shown with the model *code* TM which is explained in Annex 1. The assumptions are explained above. The study focuses on the EU proposal, which is called *EU prop strict*. Due to the Kyoto Protocol an Annex B-wide ET system exists. In the scenario *EU prop strict* the energy intensive sectors of the EU are not allowed to take part in the Annex B-wide ET. In reference to the special meaning of the FSU in this Annex B-wide ET one has to distinguish between two cases. In the first case, competitive behaviour of FSU is assumed. In the second case, the FSU supplies its Hot air strategically. This means that FSU acts as a partial monopoly and Eastern Europe is the fringe supplier. In reality, the truth can be found between the two theoretical cases.

In this analysis the second question is about the ecological and the economic impact which would be caused by a flexibilisation of instruments. In a first step it is assumed that the energy intensive sectors in EU are allowed to realise JI and CDM projects instead of a sole domestic reduction. In this scenario, which is called *EU prop wide*, the Annex B-wide ET system still exists isolated from the EU internal one. Only the complete integration of the Kyoto mechanisms leads to the cancellation of the isolation. In such a case the energy intensive sectors of the EU were allowed to take part in the Annex B-wide ET system. This second and most flexible scenario is called *Marrakech*.

It has to be indicated that the costs for JI and CDM projects result endogenously for each region and are not fixed exogenously. This means that potential CO_2 reduction projects will be offered and, depending on the world-wide CO_2 price, projects will be realised, or not. The price of a project depends highly on the endowment of the region with the necessary production factors. It is assumed that financing regions have no preferences for a host country.

The Danish proposal can be seen as an approach to make the EU proposal more flexible. But JI and CDM projects are just mentioned as desirable, and the participation of energy intensive sectors at the Annex B-wide ET system is only seen as a possibility. In contradiction to this vague formulation, the following argumentation bases on a legally binding implementation in the ET scheme.

The following scenario descriptions firstly analyse the resulting price for CO_2. In a second step the analysis focuses on the emission reductions and the leakage. By this, the first step describes the primal effect of the economic dimension and the second analyses the ecological goal. In a third step, the impact on the national energy mix is taken into account as it is an important political aim of the EU and of Germany. This third step relativizes the results on the background of a system of political goals.

5.5 Scenario *EU prop strict*

The EU proposal from the 23rd of Octobre 2001 is analysed in the scenario *EU prop strict*. Thus, the energy intensive sectors in the EU have no possibility to reduce their commitments by realising JI or CDM projects, neither are they allowed to take part at the ANNEX B-wide ET system. In fact they are isolated in the EU intern ET system.

- CO_2 price

It is obvious that an isolation of the energy intensive sectors in the EU leads to a CO_2 price which is different from that on the Annex B-wide permit market. The huge amount of Hot air available outside the EU internal market leads to a very low price. A competitive behaviour of Russia means a price of between zero and 0.4 UScent/t CO_2. Although it is not realistic that Russia will give their Hot air away for free the assumption of no market power leads to very low prices. The strategic supply behaviour is analysed in chapter 5.7 in form of a partial monopoly.

The price for CO_2 rises within the EU internal ET system sharply from 2.80 \$/t CO_2 in 2006 to 10.40 \$/t CO_2 in 2008. Figure 6 displays the price path. In this period the permits are devaluated after they have been distributed via Grandfathering in 2005. The absolute cap assigned by the Kyoto protocol is held fix between 2008 and 2012. This ensures that the commitment will be kept as an average of this Kyoto target period. The assumed growth leads to a further increase of carbon prices, which will result in a price of 12.70 \$/t CO_2 in the year 2012.

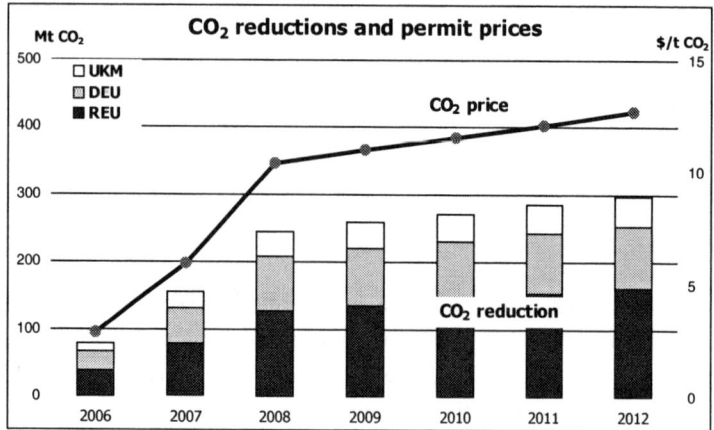

Figure 6: CO_2 reductions and permit prices if the EU internal ET system is isolated and JI and CDM projects are not allowed[45]

- Emission reductions

Most of the emission reductions will be realised in the region Rest of Europe. There CO_2 emissions will be reduced by more than 160 Mt CO_2 until 2012. The German CO_2 emissions will be reduced by 92 Mt and those of the UK by 44 Mt. These amounts are valid for the year 2012 and sum up to a EU total of nearly 300 Mt. This proves that Germany does not keep its commitment by domestic reductions and will not be a net supplier of permits. More than half of Germany's remaining gap of 180 Mt CO_2 will be bought outside. This affects the EU internal ET system which allows the energy intensive industry to buy permits in the Region Rest of Europe. The government is responsible for the non-committed part of the economy concerning households and traffic. The gap resulting from this part will be covered, according to the Kyoto protocol, by the Annex B-wide ET. Contradicting the EU forecasts, Germany will not be a net supplier of permits, which proves that reduction costs in Germany are higher than in other European regions.

The world-wide leakage will be 20.1%. The leakage is defined as the relation between the increase of emissions in non-restricted regions and the decrease in regions with a reduction commitment. The permit prices of nearly zero mean that, in fact, only the energy intensive sectors in the EU get a restriction. The competitive behaviour of Russia as a supplier of permits prevents the Rest of Annex B from a real restriction. In this case, the leakage means that of the

---

[45] The country codes are listed in Annex 1.

reductions of nearly 300 Mt CO_2 in the EU 20.1% are without ecological use, as production in other regions rises. This rise leads to an increase of CO_2 emissions by more than 60 Mio. t.

The leakage has two reasons. First, the production rises in non-restricted areas. The restriction in the EU leads to a drop within the domestic production. Competitors in non-restricted countries can enlarge their international market share by the created cost advantage. Secondly, the price effect on the market of international fossil fuels leads to a drop in prices for hard coal and oil products. Because of the homogeneity of these products the price effect is a global one and creates a tendency in non-restricted areas like the USA and the Non-Annex B-countries to use more hard coal and more mineral products.

Excursion

Supply elasticities have a significant meaning for the price effects on the international market of fossil fuels. This index stands for the change in supply caused by a change in price. Different studies have underlined the importance of supply elasticity on the international coal market. Additionally, these studies concluded that the assumption of low supply elasticities leads to a sharp rise in leakage.[46] This conclusion can be transferred to the gas market as another influential market for this discussion. The basis for the index number, besides the strategic orientation of supply and demand, are the transport and production capacities and the production costs. Sensitivity analyses of the used model *code* TM show that under extreme assumptions the leakage can switch to an ecological contradiction. In such a case the leakage can lead to an increase in global CO_2 emissions in 2012 of 2.3% if the EU proposal will be realised. This means that the leakage is higher than 100% and the ecological net effect is negative. It has to be emphasised that this negative effect only occurs in connection with the EU proposal. This shows that the central goal of climate change policy must be the reduction of global CO_2 emissions in order to minimise the leakage.

The EU proposal leads to a restriction of a small number of emitters in several regions only. In fact, for most emitters no real restrictions are set, so that for a great part of the restricted regions, the price effects of fossil fuels lead to an increase of their CO_2 emissions. This holds true for the traffic in the EU.

---

[46] Cf. Paltsev (2000), Burniaux/Martins (2000) and Manne/Richels (2000).

The competitive behaviour of FSU will set no real restriction for the Non-Annex B-regions and the non-included EU-sectors like traffic and households. For the governmental ET system a price of (nearly) zero results, while in the isolated EU internal ET system CO_2 is traded for 12.70 $/t. The huge amount of Hot air, in connection with the assumption about the supply behaviour of the FSU, leads to an increase of CO_2 emissions of about 33 Mt in the year 2012. In the whole Annex B area the CO_2 emissions will decrease by 270 Mt in 2012. The question is, if it is possible to lower the leakage to reach a higher reduction of global CO_2 emissions and in which way the conception has to be changed. Before comparing different concepts in the ecological dimension, the energy mix as another important political goal has to be analysed.

- Energy mix

To judge the EU proposal, not only the price of carbon and the impact on global emissions have to be taken into account, but as well the implementation in the wider system of political goals. In this connection the economic effects play an important role, and so they are analysed on a sectoral level in chapter 6. At this point, the on EU level defined political goal of the safety of energy supply shall be analysed. Depending on the conception of an ET system this goal will be influenced in different ways.

The basis for the question in how far a region depends on imports to insure its energy demand is the share of fossil fuel imports. Therefore it is essential to analyse the impact of an ET scheme on the shares of each fossil fuel.

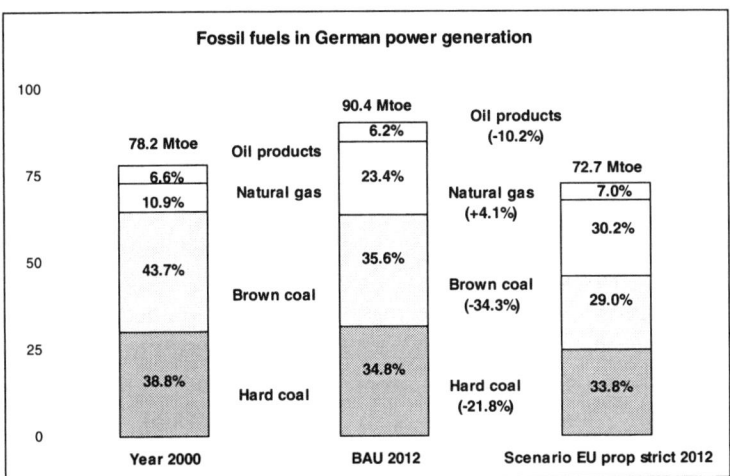

Figure 7: Share of fossil fuels in the fossil founded electricity production of Germany.[47]

The EU proposal leads to a decrease of the electricity production of 7.0%. The use of fossil fuels drops in the same time from 90.4 Mtoe to 72.7Mtoe, a decrease of 19.5%. Figure 7 displays the effect and the impact on the input of German power stations in the year 2012. The strongest effect can be stated for the input of brown coal, which drops by a rate of 34.4%. In the scenario a share of only 23.4% remains for brown coal. The production based on hard coal decreases by 21.8% and has a share in 2012 of 27.2%. The demand of natural gas rises slightly by 4.1% and covers the fossil founded production by a share of 24.3%. The use of oil products decreases by 10.2% and covers the rest of 5.6% of the German fossil founded electricity production.

The most important aspect in terms of supply security is the division concerning the origin of fossil fuels. Although the non-fossil fuels as fast growing factors are part of the domestic input the supply security will not improve. All regenerative energies depend on their respective natural sources, like sun, wind, or water, and if this source is not active, no electricity will be produced. Fossil fuels have to be analysed concerning their regional availability. Brown coal is covered by domestic production. Steam coal is imported in significant amounts. The impact of the rising CO_2 price on division between the domestic

---

[47] The data of the year 2000 are based on VIK-data, cf. VIK (2001), Table 103. Reference case and scenario values result from own calculations with the modell *code* TM.

production of hard coal and the imported one depends significantly on the assumption about hard coal subsidies.

Excursion

If every difference between the price for import coal and the domestic one is equalised by the subsidies, the absolute amount of domestic steam coal will be stable and the pressure on the demand of coal will only affect the hard coal imports. The system of yearly fixed funds founded in 1997 leads to an annual decrease of hard coal production until 2005 if subsidies shall be lowered to 2,5 Bio. €. This leads to a decrease of the domestic production. The current political discussion hints to at solution wherein a budget based system will remain. Until 2005 the EU has agreed to this system. The new regulation for hard coal subsidies passed in mid 2002 and allows further financial aids until 2010. The regulation passed for two reasons. First, the subsidies allow a restructuring of regions depending on coal production, and second, to ensure a minimum production of domestic hard coal. In the German case an amount of 20 Mt is discussed.

An additional cost pressure for the German coal production is caused if production made it necessary to buy emission permits. The biggest part of energy input is electricity. This factor would lead to a rising cost difference between the domestic production and imports due to the impact of the price of CO_2 on electricity production. This will be an advantage for regions without an emissions restriction like South Africa and Columbia.

Depending on the German subsidy system the supply safety in the year 2012 will be affected differently. According to the political discussion a budget system is assumed for the model, which leads to a cost pressure for the German coal production. A decreasing production leads to an additional rise in costs due to the *economies of scale* and a further reduction is likely. In the case of the EU proposal the budget will help to protect the domestic production which will fall only by 5.0%.

Altogether, no reliable conclusion can be drawn about the percentage of electricity founded domestically and safe. The definitive conclusion is that because of the drop in the use of brown coal the share of domestic fossil fuels decreases. Therefore, the dependence on imports from countries like Russia and Algeria will rise significantly.

5.6 Scenario *EU prop wide*

If the scenario described above is modified by an introduction of JI and CDM projects, the results change significantly.

- CO_2 price

In this case, the resulting CO_2 price is about 3.30 \$/t CO_2 in 2006. It decreases continuously until 2012 where the permit price is nearly 2.60 \$/t CO_2. The decrease is caused by the great amount of JI projects realised in East Europe where a significant part of the electricity sector will be renewed. These new production capacities lead to an annual inflow of CO_2 permits. Future reduction commitments are reduced by those permits which depend on how long a JI project reduces emissions. This reduction over a period of several years leads to a cumulative effect.

As this scenario includes also an isolation of the EU internal ET scheme from the Annex B-wide system, again two prices for CO_2 result. The Annex B-wide stays zero as it is still assumed that the FSU does not act strategically. Therefore only the EU internal ET scheme gets a price signal for CO_2 and realises JI projects (only) in Eastern Europe. The shadow prices show that slightly higher carbon prices would lead to profitable JI projects in the region FSU.

It is assumed that JI and CDM projects only contain CO_2 reducing measures in power generation. This means that no projects about carbon sinks will be allowable. In figure 8 the price path of CO_2 is displayed on the background of the regional reductions.

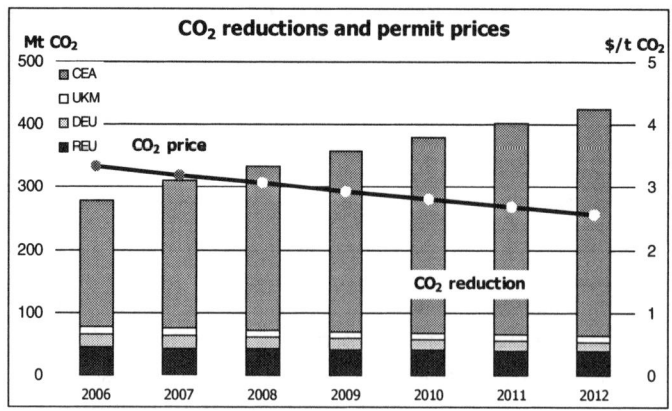

Figure 8: CO_2 reductions and permit prices with an isolated EU-wide sectoral ET scheme and with JI and CDM projects[48]

- Emission reductions

While the global emissions drop in the first scenario to 29,510 Mt CO_2, their amount decreases in this scenario by 170 Mt to 29,340 Mt CO_2. The flexible mechanisms lead not only to a lower price but also to lower global emissions.

The domestically emitted amount of CO_2 decreases in Germany by 15 Mt CO_2 in the year 2012 in comparison with the reference case. The CO_2 emissions of the UK will be reduced by 9 Mt CO_2 and those of the region Rest of Europe by more than 38 Mt CO_2. Even in this case the biggest reduction of Europe will be realised in the region Rest of Europe. This has to be stated in contradiction to the expectations of European politicians.

Production rises only in the FSU significantly. This leads to an increase in CO_2 emissions of the region FSU of 0.5%. The overall leakage will be about 13,5%. JI projects will only be realised in Eastern Europe, where CO_2 emissions drop by 38% in 2012. The resulting price of CO_2 does not lead to further JI or CDM projects in other regions. Crucial are the regional reduction costs, the predominating CO_2 efficiency, and the availability of natural gas. Specially the Non-Annex B-regions set lower incentives. Only from a price of 4.50 $/t CO_2 on CDM projects will be realised. The more countries act as host countries for JI and CDM projects, the higher will be the net effect for global warming.

---

[48] The country codes are listed in Annex 1.

- Energy mix

The impact on the energy mix is much lower in this scenario than it would be by the EU proposal. The use of fossil fuels drops from 90.4 to 87.0 Mtoe. Figure 9 displays the context and the demand of fossil fuel power stations in the year 2012. Brown coal will hold a stable share of about 35.6%. In contrast the use of seam coal decreases by 11.7% and its share drops to 30.7%. The natural gas input increases by 2.2%, therefore the share rises by half a percentage point to 23.9%. The share of mineral products decreases from 6.2% to 6.1%.

The conclusions about the security of energy supply hold true for this scenario as well. In this case, the German hard coal production drops by 5.9%. The dependence of the German energy production on imports from Russia or Algeria rises significantly, but less than in the first scenario *EU prop strict*.

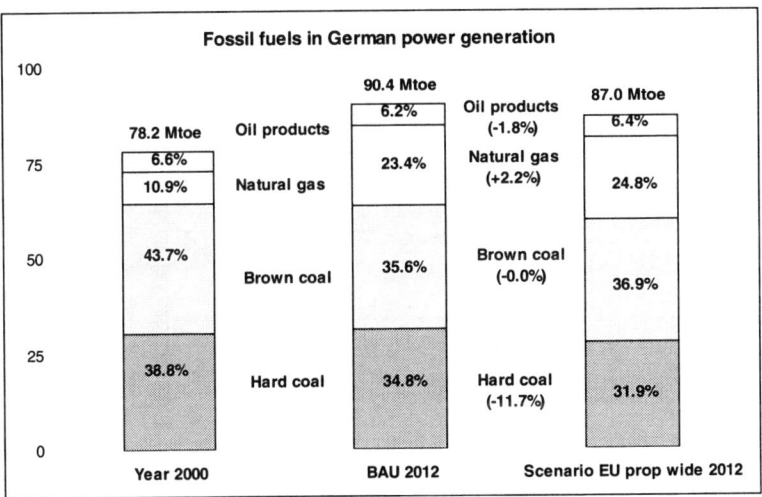

Figure 9: Share of fossil fuels in the fossil founded electricity production of Germany [49]

5.7 Scenario *Marrakech*

While the comparison between the first and the second scenario shows the positive effect of JI projects to fight global warming, the effect of a cancellation of the isolation of the EU internal ET scheme is analysed in this scenario.

---

[49] The data for the year 2012 are provided by VIK, cf. VIK (2001), Table 103. The reference case and the scenario results are provided by the model *code* TM.

In contrast to the EU proposal, it is assumed that energy intensive sectors use the mechanisms which were agreed upon in the Marrakech accords. This means that a company bound by the EU proposal will be able to take part in the Annex B-wide ET scheme. Institutionally this can be realised by a governmental person who hands the permits down by charging the price in the Annex B-wide ET system. Additionally, it is assumed that companies will be able to realise JI and CDM projects.

- CO_2 price

In contrast to the scenario *EU prop strict*, the demand of the energy intensive industry appears on the Annex B-wide permit market. Thus a positive price for CO_2 results. In 2012 the price will be about 0.33 \$/t CO_2. Figure 10 displays the price path on the background of the regional emission reductions. This constellation leads to a total consumption of Hot air.

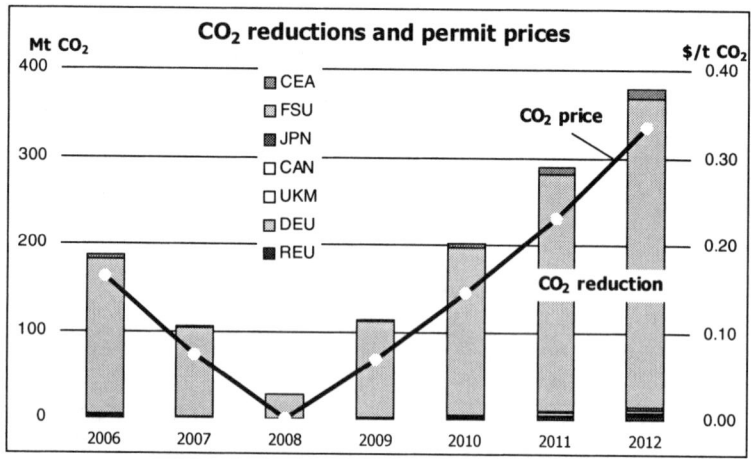

Figure 10: CO_2-reductions and permit prices under isolation of the EU-wide sectoral emissions trading and with JI/CDM-projects[50]

The path clarifies the effect of Hot air depletion. Around 2018 the whole Hot air amount of the FSU and East Europe will be consumed. Until this year the price path of CO_2 will rise superproportional.

- Emission reductions

---

[50] The country codes are listed in Annex 1.

The low price of CO_2 leads to an unexpected positive ecological effect. The global emissions drop by 1.2% and 4.7% result as the leakage. This means that the global emissions in the year 2012 will be around 29,380 Mt. The price for carbon which is significantly positive for this case leads to additional reductions specially in the region FSU. These reductions, the so called *no regret*-measures, decrease the FSU emissions of 2012 by 14.9% in comparison to the reference case. The CO_2 emissions in East Europe will be lowered by 1.1%. Canada and Australia will reduce their CO_2 emissions in the year 2012 by 0.5%, Japan by 0.3%, and each European region by 0.2%.

Additional reductions of global CO_2 emissions do not appear because of two reasons: First, part of JI based reduction will be replaced by Hot air; second, the Hot air revenues allow higher consumption paths in the FSU and in East Europe which leads to higher emissions.

The low price for CO_2 prevents a significant effect on the production as well as on the world market for fossil fuels.

- Energy mix

The consumption of fossil fuels in the year 2012 decreases in this scenario by 0.6% compared to the reference case. In this case brown coal shows the strongest effect with a decrease of 1.1%. The share of brown coal in the fossil fuel based power generation remains at 35.2%. Figure 11 clarifies the context.

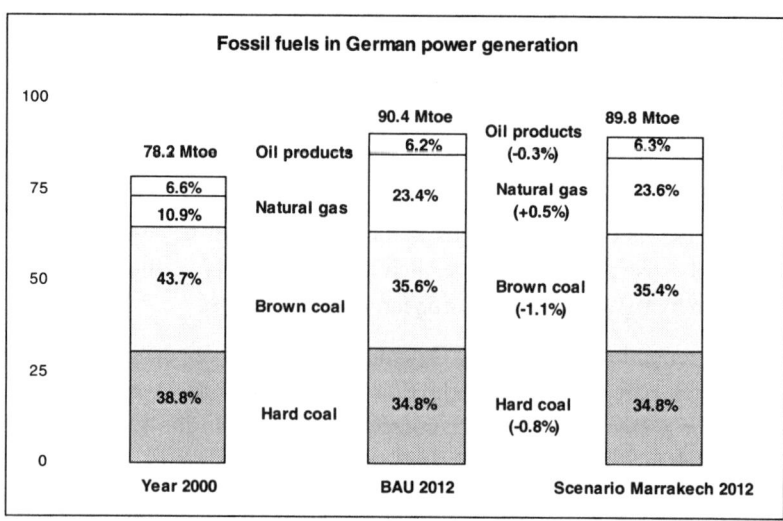

Figure 11: Share of fossil fuels in the fossil based electricity production of Germany [51]

The share of hard coal remains unchanged at 34.5% while the consumption decreases slightly by 0.8%. The increase in natural gas of 0.5% leads to a slight rise in its share to 23.5%.

Altogether, the security of energy supply is affected only slightly. An important aspect in this argumentation is that the German hard coal production decreases by only 3.6%.

5.8 Strategic supply behaviour of Hot air

In the previous scenarios we assumed a competitive behaviour of the FSU. This leads to very low carbon prices as shown above. This market form cannot be assumed as a realistic one because the constellation allows the FSU to use their market power. The market form has to be rated as a partial monopoly with the FSU as the monopolist and Eastern Europe as the fringe supplier. Eastern Europe has just a small amount of Hot air and has therefore no market power. In such a constellation the monopolist tries to realise the Cournot price.

The theoretical model says that the monopolist sets the Cournot price on basis of the rest of the demand. This means the total demand has to be reduced by the fringe supply. By this, the amount on which the Cournot price is founded

---

[51] The data for the year 2012 are provided by VIK, cf. VIK (2001), Table 103. The reference case and the scenario results are provided by the model *code* TM.

results. In this optimisation process the possibility to realise JI and CDM projects sets an upper limit for the carbon price. Therefore the demand curve is a deviated one. Above the breakpoint the FSU will not be able to sell any more Hot air because emission reductions from JI and CDM projects will be cheaper. On this mechanism the scenarios defined above will be analysed in connection with a strategic supply behaviour of the FSU. To analyse the two market forms (competitive and strategic behaviour) allows the definition of a lower and an upper limit of the forecasts.

Crucial for this behaviour is that FSU itself is part of the group supplying JI projects. The FSU optimisation process implements the question which aspect has a better impact on utility:

- The realisation of JI projects to improve the domestic efficiency with foreign capital, or
- the current import by selling emission permits.

As an additional bound exists the role of the FSU as an important supplier of natural gas. This optimisation leads to a strategic price of CO_2 of about 3.80 \$/t.[52] This price prevents further JI and CDM projects in other countries and maximises the utility of the region FSU. The break even point for CDM projects depends on the *host* region and is between 4 and 6 \$/t CO_2. The price of 3.80 \$/t CO_2 maximises the utility of the FSU by the supply of Hot air and the realisation of JI projects. This strategic price is only valid for the Annex B-wide ET scheme. In case of the EU proposal the EU energy intensive industries are still isolated. As the carbon price in the Annex B-wide ET system has a significant effect on the world market prices of fossil fuels, the carbon price in the EU internal scheme is affected indirectly.

- Scenario *EU prop strict*

Figure 12 clarifies that the price for CO_2 in the year 2012 will be around 15.30 \$/t, while it was about 12.70 \$/t in the case of competitive behaviour. This change occurs although the restriction for the energy intensive sectors in the EU remains unchanged. Just because of the strategic price of 3.80 \$/t CO_2 the situation of the energy intensive sectory declines.

---

[52] For a forecast made step by step see Böhringer (2002), pp 65

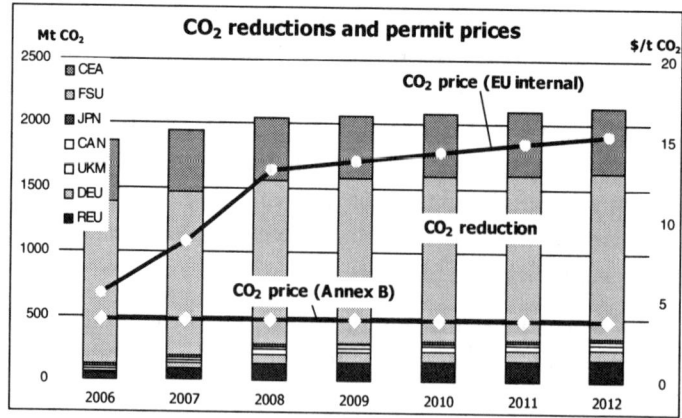

Figure 12: CO_2 reductions and permit prices in an isolated EU internal sectoral scheme without JI and CDM projects, with a strategic supply behaviour of the region FSU[53]

The world-wide emitted amount of CO_2 emissions will be reduced by 4.7% compared to the reference case. This shows how important a significant restriction of Non-EU regions is for global warming. The question about how far the instrument Emissions Trading is implemented (number of countries) seems to be more important than the question about the deepness (reduction commitment) as long as the restriction works. Therefore it is more efficient to implement more installations and countries with low commitments than only a few with high reduction targets. This insight can be transferred to the leakage problem. For this scenario results a leakage of 34.4%. The strategic price for CO_2 of 3.80 $/t leads to a huge amount of JI projects in the regions FSU and Eastern Europe.

The energy mix changes significantly. The use of fossil fuels in German power generation is reduced by 18.4%. The input of brown coal drops by 27.7% and its share sinks to 25.7%. The share of hard coal decreases to 26.8% while that of natural gas remains nearly unchanged. Figure 13 displays the shares each fossil fuel holds within the total demand of German fossil fuel power stations. It can be concluded that the security of energy supply is affected significantly.

---

[53] The country codes are listed in Annex 1.

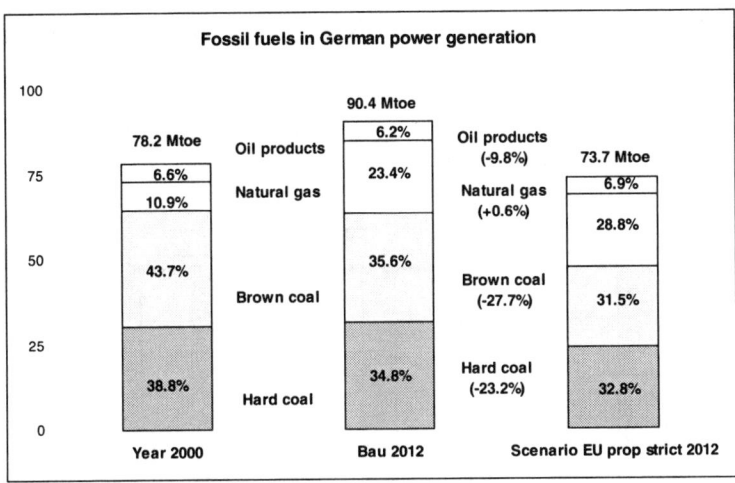

Figure 13: Share of fossil fuels in the fossil founded electricity production of Germany [54]

- Scenario *EU prop wide*

In the case of an integration of JI and CDM in the EU proposal on the level of companies the carbon price will be equalised for both ET schemes, the EU internal one and the Annex B-wide one. In case of the energy intensive industries isolated in the EU internal system, the realisation of JI projects is significantly cheaper than domestic reductions. As the strategic price set by the region FSU is higher than the price for JI projects the isolation is indirectly annulled. The cheapest option for the energy intensive sectors of the EU equals the lower limit in the Annex B-wide ET system. The price of CO_2 in the EU-internal ET scheme will increase to 3.80 $/t while it was about 2.70 $/t in the case of competitive behaviour. The rise in the Annex B-wide ET system is significantly higher, from nearly zero to 3.80 $/t CO_2.

Depending on the assumptions of JI projects, a part of the permit demand is covered by cheap CO_2 reduction projects in the region Eastern Europe. The supply side of Eastern European JI projects is defined by a disunited group of many members. Thus they have no market power and FSU is able to set its strategic price although the price of Eastern European JI projects is significantly lower. The Eastern European suppliers of JI projects act as price takers

---

[54] The data for the year 2012 is provided by VIK, cf. VIK (2001), Table 103. The reference case and the scenario results are provided by the model *code* TM.

and sell their JI contingents for 3.80 $/t CO$_2$ instead of the former 2.70 $ t/CO$_2$.

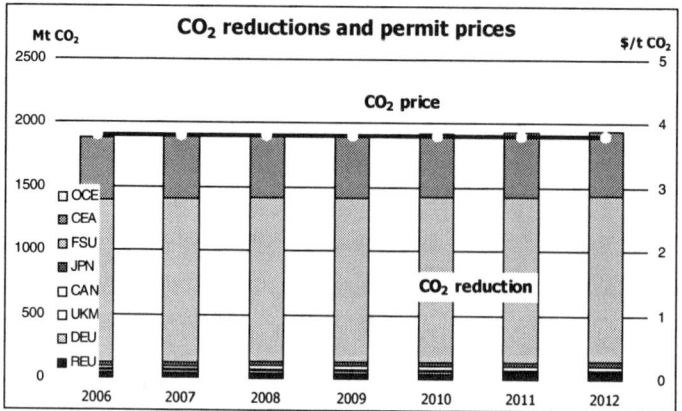

Figure 14: CO$_2$ reductions and permit prices in an isolated EU internal sectoral scheme with JI and CDM projects, and with a strategic supply behaviour of the region FSU [55]

As displayed in figure 14 nearly all global CO$_2$ reductions will be realised in the regions FSU and Eastern Europe. Global CO$_2$ emissions of the year 2012 drop by 5.9% compared to the reference case and will be around 28,000 t. For the JI mechanism the ET scheme is more flexible and the leakage will be 10.9%.

---

[55] The country codes are listed in Annex 1.

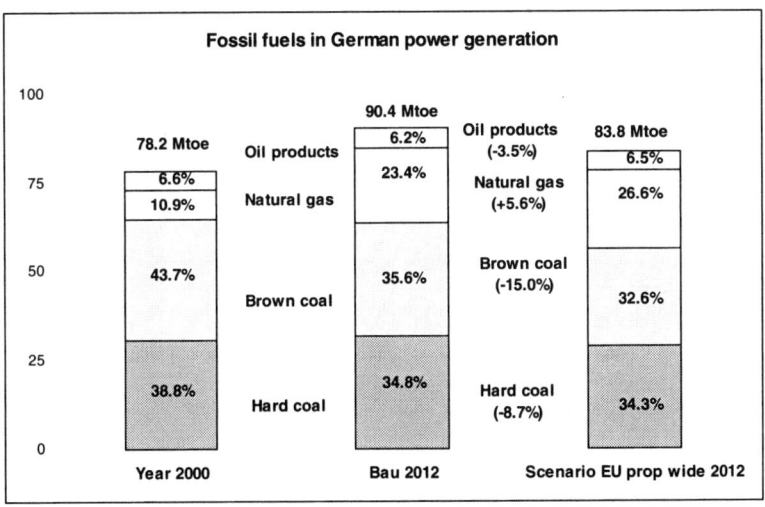

Figure 15: Share of fossil fuels in the fossil founded electricity production of Germany [56]

The impact of such a concept on the German energy mix is displayed in figure 15. The share of brown and hard coal decreases while the share of natural gas slightly increases to 24.7%. The total amount of fossil fuels used in fossil fuel power plants drops by 7.3% in comparison to the reference case.

- Scenario *Marrakech*

The scenario describes the direct annulment of the so far isolated ET scheme of energy intensive industries in the EU. As the prices for CO_2 were equalised for both ET schemes by the implementation of JI projects and the strategic behaviour of the FSU in the scenario *EU prop wide* the results remain unchanged. Thus the same price of 3.80 $/t for CO_2 results, which is the strategic price of the region FSU. The global emission reductions again lie, slightly lower than in the scenario *EU prop wide*, at about 5.9% with a total amount of CO_2 emissions of 27,990 Mio. t CO_2.

---

[56] The data for the year 2012 are provided by VIK, cf. VIK (2001), Table 103. The reference case and the scenario results are provided by the model *code* TM.

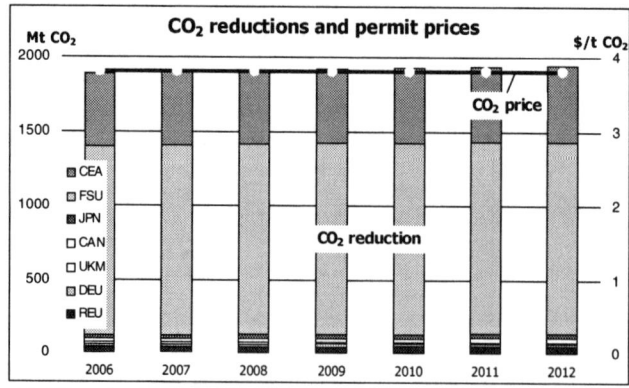

Figure 16: CO_2 reductions and permit prices in an Annex B-wide ET scheme with JI and CDM

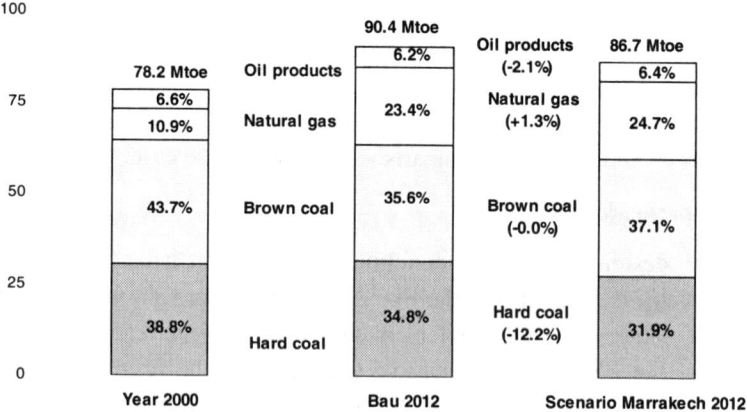

projects, with a strategic supply behaviour of the region FSU [57]

Figure 17: Share of fossil fuels in the fossil founded electricity production of Germany [58]

The impact on the German energy mix is shown in figure 17. The share of brown coal at the total amount of fossil fuels in power generation remains nearly constant. The share of hard coal in the year 2012 decreases to 30.6%, 4.2 percentage points less than in the reference case. The share of natural gas

---

[57] The country codes are listed in Annex 1.
[58] The data for the year 2012 is provided by VIK, cf. VIK (2001), Table 103. The reference case and the scenario results are provided by the model *code* TM.

rises slightly to 23.7%. Compared to the reference case, the total amount of fossil fuels sinks by 4.1% lower than in he reference case.

5.9 Concluded model results

In comparison to the EU model PRIMES the carbon prices of the model *code* TM used in this study are lower. The EU proposal is analysed in the scenario *EU prop strict* and defines the isolated EU internal ET scheme for energy intensive industries. These sectors are not allowed to participate at the Annex B-wide ET system nor are they allowed to lower their own commitments by realising JI or CDM projects. The price of CO_2 in such a scenaio is just half of that one forecasted by the PRIMES study (30 €/t CO_2). The difference is caused specially by one factor: PRIMES includes solely the EU member states as active regions. The other regions of the world are not implemented endogenously. Thus, the EU model is not able to forecast leakage effects as production locations outside of the EU cannot be shown. In addition, the assumptions within this model exclude important price effects on the world-wide market, specially for fossil fuels.

Scenarios are described in the following way: In a first step the EU proposal is analysed, and is displayed by column 3. Energy intensive sectors of the EU are allowed to trade only within the isolated EU internal ET system. In a second step the scenario *EU prop wide* is analysed, which is enriched by the possibility for the energy intensive countries in the EU to realise JI and CDM projects, as displayed in column 4. The next step of flexibilisation is the possibility to participate in the Annex B-wide ET system, column 5. In the last scenario exists just one price of CO_2 world-wide. With this structure it is possible to analyse the ecological and economic impact of an increasing flexibilisation of the ET scheme.

Table 4 displays the results if it is assumed that the region FSU supplies its amount of Hot air competitively. The most important index numbers are shown. The first step of flexibilisation (from column 3 to column 4) shows the impact of introduced JI and CDM projects: The price of CO_2 in the EU internal ET system drops from 12.70 $/t to 2.60 $/t. At the same time global emissions will be significantly lower, a reduction of 1.3% results while the reduction in the scenario *EU prop strict* was only 0.8%.

(1)	(2)	(3)	(4)	(5)
2012	Reference case	**EU prop**		Marrakech
		strict without JI/CDM	wide with JI/CDM	with JI/CDM and with Annex B-wide ET
Global emissions (Mt CO_2)	29,740	**29,510**	29,340	29,380
change of global emissions compared to reference case		**-0.8%**	-1.3%	-1.2%
leakage effect		**20.1%**	6.0%	4.7%
absolute CO_2 reduction in the EU (Mt CO_2)		**300**	63	8
absolute CO_2 reduction in all restricted regions (Mt CO_2)		**300**	425	380
absolute CO_2 increase in regions without a restriction (Mt CO_2)		**62**	25	18
CO_2 price EU internal ET system Annex B-wide ET system	-	**12.70 $/t** **0.00 $/t**	2.60 $/t 0.00 $/t	0.33 $/t 0.33 $/t

Table 4: Results without a strategic supply behaviour of the region FSU

In a second step, the impact of an implementation of the EU Emissions Trading scheme in the Annex B-wide system, according to the Marrakech accords, was analysed. By this implementation just one carbon price results, 0.33 $/t in the year 2012. At the same time, global emissions rise slightly in comparison to the second scenario. In a longer term the net effect would be much more positive as the amount of Hot air would have been consumed from 2018 on and more JI and CDM projects would have been realised. In comparison to the reference case, in the year 2012 there will be a global reduction of about 1.2%.

In a case of a strategic supply behaviour of the region FSU a lower limit for the price of CO_2 will be set at 3.80 $/t. Table 5 displays the most important results.

(1) 2012	(2) Reference case	(3) EU-RLV		(4)	(5) Marrakech with JI/CDM and with Annex B-wide ET
		without JI/CDM	with JI/CDM		
Global emissions (Mt CO_2)	29,740	28,340	28,000		27,990
change of global emissions compared to reference case		-4.7%	-5.9%		-5.9%
leakage effect		34.4%	10.9%		10.9%
absolute CO_2 reductions in the EU (Mt CO_2)		295	75		75
absolute CO_2 reductions in all restricted regions (Mt CO_2)		2145	1930		1930
absolute CO_2 increase in regions without a restriction (Mt CO_2)		740	215		215
CO_2 price EU internal ET system Annex B-wide ET system	- -	15.30 $/t 3.80 $/t	3.80 $/t 3.80 $/t		3.80 $/t 3.80 $/t

Table 5: Results with a strategic supply behaviour of the region FSU

The strategic supply of Hot air of the region FSU leads in connection with the EU proposal to a global reduction of CO_2 emissions of 4.7%. The additional implementation of JI and CDM projects in this case also has a positive impact on the global reductions. A reduction of 5.9% will be reached in the year 2012. The reduction has the scenario of Marrakech which allows the Annex B-wide ET due to the same strategic price of CO_2.

The assumption about the strategic option of the region FSU has to be put into perspective. The demand on the permit market will have strong incentives to prevent the Cournot price. In international politics several factors will allow the contradiction of a lower limit for permits. Therefore the sure lower limit of the CO_2 price of 3.80 $/t is, in reality, unsure.

The results of the AGE model will be used as the needed input for the sectoral optimisation model of the RWI. Beside the EU forecast of 30 €/t CO_2 the scenario *EU prop strict* (about 15 €/t CO_2) from the *code* TM forecast will be taken. Additionally, the further context will be simplified by summing up the scenarios *EU prop wide* and *Marrakech* with a price of 5 €/t CO_2. This price is justified due to implications about the endogenous costs of JI projects. This price can also be interpreted as an upper limit of sectoral effects.

5.10 Model variations and the interpretation of results

5.10.1 The sectoral aggregation

The CO_2 prices are the direct price impulse for the analysed impacts. The increase in the price of electricity can be defined as an indirect impulse. Reality can be implemented in a model in a very simplified structure. Therefore, the comparison of real conditions and a given model structure has a significant meaning. Specially regarding the question in how far the abstraction of important aspects affect the direct and the indirect price effects.

An important aspect in evaluating model results is the model aggregation. The sectoral aggregation listed in Annex 1 will be changed in this chapter to analyse the impact of a change in the aggregation. We will use the metal sector (ISM) for this sensitivity analysis. The analysis above aggregated non-Iron metal, Iron and steel, in one sector. Steel production is not differentiated in terms of different production plants. In the following, a differentiated view shall allow an insight in the impact of sectoral aggregations. Obviously the leakage effect is a key mechanism within the climate change discussion. This implements the question how easy domestic products can be replaced by imports and therefore how homogeneous products are modelled. Oxygen steel, for instance, is in real terms widely a homogeneous product as well as electro steel is. The heterogeneous way to implement these two kinds of steel creates, within the model, an artificial protection of domestic steel production. But the substitution for the demand side is very easy because there are no preferences, only the price on the world market is a significant indicator. This competition has to be implemented in a model.

Steel will be under high pressure of international competition if homogeneity is assumed. This realistic condition leads to a price barrier of about 5-7 $/t CO_2 for the steel sector production in Europe. A price of CO_2 which is higher will lead to a shut down of the European steel production. This includes the production of electro and oxygen steel. Such a price of CO_2 will equalise the transportation costs for steel from other regions like Latin America (specially Brazil), the USA, North Africa, and some Asian countries. Due to the lack of preferences on the demand side, the origin is not important and the market share of European steel industry will be taken over by regions without an emissions restriction.

In reality the shift of production can mean a real transfer of production capacities of the same owner or the lost of market shares to competitors. This production effect will decrease the demand on the permit market of those companies with high reduction costs. The price of CO_2 as the direct price impulse will decrease and the logic context of leakage and the price of CO_2 is clarified: **The higher leakage under the same assumptions is, the lower will be the price of CO_2.**

The argument of re-locations is not meant in terms of a sudden shut down of all steel production plants. The installations will be used on the basis of middle and long term investment plans. Reinvestments or new investments will not be realised in the EU but in regions without a carbon restriction. This means that production in European steel plants will be faded out within 10 to 15 years after an agreement about a carbon emissions restriction with absolute caps on installation level as it is planned in the EU proposal. **The EU will export their carbon emissions and import the produced steel products.** Regions like South America, especially Brazil, North Africa, and the USA will profit from such a production shift. Important within this context are the transportation costs, which, in a first step, will bring mainly to those sites with connection to shipping traffic, high advantages in international competition.

Similar effects will occur in other sectors which are modelled as heterogeneous aggregated goods. This implements the glass and the cement production, which are aggregated for statistic reasons. Three quarters of the glass production is about homogenous goods and are therefore confronted directly in the international competition. As was already shown for the steel industry, a price limit exists as well. If the carbon price exceeds this limit the production capacities will be shifted out of the EU. The period for such a process in the glass industry is about 8 to 10 years. To clarify one important aspect: This argument is not an argument against every kind of ET, but concepts like the EU proposal will lead directly to production shifting and the decrease of employment.

5.10.2 CO_2 emissions caused by raw materials

The analysis does not include not all carbon emissions implemented in the EU proposal. Only carbon emissions caused by energetic use are restricted in the model, while emissions caused by raw materials are kept out of the calculation. CO_2 emissions caused by raw materials submit unavoidable emissions in the transformation process of mineral carbonates, i.e. deacidity or melting of lime stone and the input of carbonates as reduction material in the smelt of iron. This aspect mainly concerns the production of steel, cement, glass, and lime.

The integration of this kind of CO_2 emissions would increase the direct price impulse significantly. At the same time the indirect price effects will be boosted. In the case of the example of steel production this implies that through the rising demand on the permit market, the price of CO_2 will increase. The prices of electricity, lime and fire-proof material will rise as the indirect price effect. As 60% of the overall CO_2 emissions in the steel sector respectively 85% in an smelting plant for the production of oxygen steel are caused by the input of raw materials, the integration of this kind of carbon emissions will decrease the price limit for plant decisions significantly. Similar contexts can be analysed in the production of glass, cement, and lime.

A concept for an Emissions Trading scheme shall install incentives to increase the CO_2 efficiency. **Due to the existence of physical barriers, the only possibility is tocut down the production to reduce CO_2 emissions caused by raw material input.**

5.10.3 *System efficiency of co-ordinated production plants*

Modelling an ET scheme always means to work with average data. The prices for CO_2 are a result of the optimal behaviour of the modelled average producer. The producer with the lowest marginal reduction costs will cut emissions first. An above-average CO_2 efficiency is an advantage in such a scheme, the primal idea of the theoretical concept.

This way of modelling is problematic if the CO_2 efficiency depends on another limit of the system as that taken as a base in the ET scheme. The EU proposal defines the level of installations as the base for CO_2 restrictions. In many sectors as the chemical industry, steel production and production of glass, energetic processes are optimised in very complex co-ordinated production plants. This allows the cost minimisation by a given lower limit of quality. The EU proposal contradicts this co-ordinated plant concept of production and energy production. Incentives are set to cut out single installations, for instance steam production for power production outside the own system. The optimised use of couple energies will be given up for incentives of the EU proposal. This concerns specially the chemical industry and power generators. But also co-ordinated production plants like the chain coking plant-powder metal facility-furnace-steel mill, including power plant in an oxygen steel production plant. The latter one use coke coal gases, which were set free by the input of raw materials, in the energetic co-ordination.

Such an optimised co-ordinated production plant concept is, from the perspective of global warming, not only desirable but also necessary. Therefore, the focus on installations in the EU proposal has to be reviewed. The production concepts to optimise energetic systems make it necessary to choose co-ordinated production plants or locations as a basis for emissions restrictions.

6 Sectoral and macro economic impacts of the EU proposal

6.1 Preliminary remarks

The introduction of emission permits is connected with a significant change in treating property of the formerly collectively used atmosphere as a dump for Greenhouse Gases. This change from a public good to a tradable property right has a deep impact on the economic system. Additionally, the EU proposal leads to a change for just a part of the economic system and the impact is concentrated on a few member states of the EU.[59] Thus, the impact will not be limited on the demand and the production of energy. Differences in the carbon intensity of production and demand processes and the resulting increase in costs will have an impact on the competitiveness of single products or product groups. This effect will have an impact on macro economic dimensions like the growth rate, the inflation rate, and employment. This context will be analysed in this chapter. The focus lies on the German perspective, but the impact on the other member states will be analysed as well. The view on other member states is important if the differences in production and power generation structures lead to a change in relative prices in each member state, who in themselves effect the competitiveness within the EU and towards Non-EU countries.

The method used in this chapter is significantly different from that used in chapter 5. The results will not be gained by an AGE model but by a specific sectoral analysis. These results will be integrated in an Input-Output framework, respectively the national accounting. The ET scheme itself will not be discussed in this part, specially the carbon price will be set exogenously on the basis of chapter 5. The analytical consistency is not ensured by this way of modelling but the results of this sectoral approach were taken in an aggregated form as an input for the AGE model, and vice versa. The needed consistency for the sectoral analysis between the production and the consumption of goods and, on the other hand, the cost and revenue structures cannot always be guaranteed for each single sector due to the uncompleted data basis. The conven-

---

[59] This problem can be explained by a rough calulation: Assuming an EU-wide reduction of 300 Mt in the year 2012 for the group of installations defined in the EU proposal, additional costs of 9.75 Bill. € result by a proe for CO_2 of 32.50 €/t. Compared with the GDP of the EU of about 9,000 Bill. € this effect (0.1%) is neglectible. Concerning power generation in the EU of 2,500 TWh the effect means an increase of 3.9 €/MWh, 20% of the current price for base load elctricity. If the additional costs will affect only the German power generation the consequence of 20 €/MWh will be substantial.

tional Input-Output analysis bases on a sectoral aggregation which is still too rough to displays the exact cost effects in single production processes.

To allow a quantification of the impact of the EU proposal the sectoral and macro economic effects will be compared with a reference path. The reference case is defined in this step of the analysis as the ET scheme with implemented JI and CDM. According to the results of chapter 5, such a system leads to a price of CO_2 of about 5 €/t. This reference case will be compared with the potential impact of the EU proposal forecasted by the European Commission. This forecast leads to a price for CO_2 of about 30 €/t without a harm to location decisions of the energy intensive industry. In both cases the CO_2 restriction is fulfilled.

The sectoral and macro economic effects are not only caused by the carbon price but also depend significantly on the initial allocation. The EU proposal codifies an initial allocation of permits for free until 2007. Amendments in the Greenbook about the Emissions Trading scheme of the European Commission and the political discussion in the EU parliament make it likely that a part of the contingent will be allocated by auctioning in the second period, from 2008 to 2012. Therefore we will assume a partial auctioning for 20% of the contingent in the period from 2008 to 2012. Additionally, we will assume consistently that the price of CO_2 will be 5 €/t respectively 30 €/t. To allow an analytical division of the impact of the initial allocation from the cost and price impulses of the ET system itself, we will display both effects separately for the reference case as well as for the scenario. Neither the EU proposal nor the statement of the parliament contains clear information about the use of auctioning revenues. Therefore we will abstain from defining a re-allocation system.

The analysis of sectoral and macro economic impacts on an ET scheme have to be differentiated concerning short-term and long-term effects. Short-term effects describe how flexible an economic system can react towards a change in relative prices with an increase of efficiency and substitution effects in an invariable capital stock. The flexibility in the energy intensive industry is limited and the relation between energy and capital is rather complementary in a short-term perspective. The impact of an ET scheme will mainly occur in a long-term process. This process will describe a re-structuring of the capital stock including possible re-location decisions of production plants. The duration of such processes depends significantly on the lifetime of production plants and cannot be limited to a special year. The installations defined by the EU proposal concern the energy sector (electricity and heat, refining, coking plants), the iron and steel production, the mineral industry (cement, lime, glass, ceramics), and the paper and pulp industry. These installations can reach a lifetime of about 40 years. Effects will not be restricted to the period until 2012. This aspect has to be considered at least qualitatively in the following analysis.

In this context the impact of an EU-wide ET on the competitiveness of single energy intensive sectors towards suppliers from Non-EU countries has a special meaning. If the EU proposal harms competitiveness and re-locations become likely, the supply and demand on the permit market will be affected. As shown in chapter 5 location shifts will lower carbon prices while all other assumptions are kept unchanged. It has to be emphasised, that re-locations will not be the consequence of the carbon price effect but the cause. They will occur earlier and, therefore, cannot be undone afterwards.

The structure of the following chapter is based on the probable direction of impacts in an ET system. This means that reactions in the energy transformation sector will be analysed at first. This leads to price effects for each demand group and, in the next step, to the change of the level and the structure of the structure of energy consumption. By the multiplication of energy prices and the consumed quantities one gets the additional energy costs of the industry respectively of the households. This cost effect itself will increase the production costs and will lead to the overall sectoral and macro economic effects. This methodical structure builds the basis of this chapter: After a description of the most important exogenous data, which will stay unchanged for all scenarios, the additional cost effects will be analysed. First, the focus will be on the energy sector, specially the power generation, secondly, the impacts on the industry will be analysed. Subsequently, the resulting sectoral and macro economic effects will be discussed. At the end of this chapter the most important results of this sectoral view will be concluded.

6.2 Framework and exogenous defaults

The cost and price effects of an ET system do not only depend on its specific concept but are also significantly determined by independent factors. These factors describe for instance the demographic development, changes in legal and institutional conditions, and influences from foreign trade and payments. It will be assumed that all these exogenous factors will remain unchanged for all scenarios, although feedback effects caused by each ET concept cannot be excluded. Specially the development of international trade patterns, exchange rates, and real interest rates will not be affected by the ET system. The same holds true for collective agreements and the negotiated net wages which remain unchanged. Rising wage claims caused by an increase in energy prices will be neglected.

A special problem within this context is described by the existence of national and EU-wide measures to fight global warming. The ideal model displays a price which contains all informations needed for investment and consumption decisions. Companies and households will found their optimisation on the price signal and the reduction commitment will be reached at minimal costs. Obviously, this price signal of emission permits makes other instruments, like laws and taxes, redundant. This holds true for German measures taken in order to minimise CO_2 emissions as well. Yet, it is rather naive to assume that after the introduction of an ET system all other instruments will be terminated by law. It is more realistic, that legal and institutional frameworks remain unchanged, independent of the fact whether the ET is restricted to a few isolated sectors in the EU or whether all sectors are included and the possibility to realise JI and CDM projects is given. Particularly, all climate change measures which were decided in the national and EU-wide context remain active. Part of this group of measures in Germany is

– the Renewable Energy Law - EEG from 29.03.2000,

– the combined heat and power cycle law (CHP) from 19.03.2002,

– the law to introduce and to develop a green tax reform, which defines an increase of taxes on fuels and electricity until 2003.

The additional costs can be ascribed to the ET scheme are caused by the reduction costs of the single measure. The substitution of carbon rich energies for fuels poor in carbon or carbon free technologies plays a key role in this context, the costs of which significantly depend on the relative energy prices. In principle, the distinction has to be made between domestic and imported fossil fuels. Prices of imported energy reflect the supply and demand conditions on the world energy markets. The domestic prices of fossil fuels display the production costs. Additionally, domestic prices for natural gas and electricity reflect the impact of the competitive opening of the European gas and power market.

Fossil fuel	unity	2000	2005	2010	2020
Crude oil					
real[2]	$/barrel	29.50	25	26	27
nominal	$/barrel	29.40	30	34	42
Hard coal imports					
real[2]	$/t	37	38	38	39
nominal	$/t	37.75	45	50	61
Natural gas					
real[2]	€/MWh	10	11	12	13
nominal	€/MWh	10.25	13	14	18
Brown coal					
real[2]	€/t	11	11	11	11
nominal	€/t	11	12	13	16

Own calculations – [1]Without excise taxes, trade and transport and without VAT; [2]prices of 1995.

Table 6: Price path[1] of fossil fuels in Germany, 2000 to 2020

World markets for fossil fuels will not cause significant changes in the current level of energy prices. The market seems to be stable, the demand and supply side quite balanced. Consequently, the price for crude oil will not exceed the current level and will be around 26$/barrel in the year 2010 in real terms, cf. table 6. Political conflicts may interfere with this stable situation in some years. With the expanding demand of some developing countries like China rising prices for crude oil are likely after 2010. Therefore the real price for crude oil will reach 27 $/barrel in the year 2020.

In principle, prices for the domestically produced fossil fuels can change independently from the prices on the world market. Admittedly, the possibility for autonomous price settings must exist, either caused by lower production costs or monopolistic market structures. If at all, low production costs can only be stated for brown coal in Germany. While for the German hard coal production this condition is not valid any more for over 25 years, the competitiveness of brown coal was obtained by deep cost reduction programs which prevented the loss of competitiveness regarding the cheap import of hard coal. Additionally, the liberalisation of the electricity markets caused a competition about the costs of power generation not as had been the case so far about the prices for fossil fuels. Therefore, brown coal can only hold its competitiveness towards the import of hard coal if the power plant price for brown coal allows a competitive power generation compared to imported hard coal. This is only possible if the price for brown coal remains stable compared to the import price for hard coal, which means that its real price remains constant in the next 10 to 15 years.

An important meaning for the following analysis has the price path of natural gas due to its low content of carbon. So far, the price for natural gas based on viability was orientated at the price of other fossil fuels with additional fees for its easy handling or the environmental advantages. It is likely that this mechanism cannot be held in a competitively organised market. On the other hand is it not possible to find the prices for aural gas solely witzh the help of marginal production costs, as the biggest part of the gas prices consists of the costs for transportation and distribution nets, which remain indispensable in a competitive market structure. Additionally, a competition on the level of producers is unlikely as foreign suppliers deliver more than 80% of the German natural gas demand and as this foreign supply increasingly concentrates on two countries, Norway and Russia. Further on, imports of natural gas are specified by so called take or pay contracts to make the high risks bearable which are caused by the high investment in exploration activities and the construction of pipeline networks. As the European, specially the Eastern European, demand of natural gas is likely to rise significantly within the next ten years additional investment in exploration and transport is needed and long-term contracts will also remain important in the far future. On this background a decrease of prices for natural gas as they could be observed on the electricity market within the last years are unlikely.

The German costs of an EU-wide ET system for CO_2 emissions do not only depend on the change in German energy prices but also on the price effect in other EU member states. The substitution of hard coal for natural gas seems to be stronger the less prices differ between hard and brown coal on the one hand, and hard coal and natural gas on the other hand. Some other member states are in a similar situation as Germany, specially countries with a high share of domestic fossil fuels in the energy production which were explored for competitive prices. Member states like Greek and Spain, with own brown coal reserves, and Finland and Ireland, with their input of domestic peat in the energy production. The Spanish strip mining of brown coal will be faded out within the next years and will notb have a meaning for the first restriction period 2008 to 2012. But in Greek brown coal will be a stronghold in their energy production in the future as it can be produced at quite low costs. The production costs are currently about 28 €/tce, 25% lower than in Germany. This difference will remain stable in the next years, see table 7, due to the good geological conditions and the higher energy content of the Greek brown coal.[60]

	Unity	2000	2005	2010	2020
		Brown coal			
Germany	€/tce	39	46	50	64
Greece	€/tce	28	31	35	45
Finland[1]	€/tce	26	29	33	43
Ireland[1]	€/tce	32	35	39	49
Spain	€/tce	33	36	-	-
		Imported hard coal			
Belgium	€/tce	50	58	63	81
Denmark	€/tce	37	43	47	60
Germany	€/tce	42	49	53	68
France	€/tce	49	57	62	79
Italy	€/tce	53	62	67	86
Netherlands	€/tce	42	49	53	68
Spain	€/tce	43	50	54	70
UK	€/tce	52	61	66	84
		Imported natural gas			
Belgium	€/MWh	9.8	12.5	13.4	17.3
Germany	€/MWh	10.3	13.1	14.0	18.1
France	€/MWh	10.5	13.4	14.3	18.5
Greece	€/MWh	8.7	11.1	11.9	15.4
Portugal	€/MWh	10.2	13.0	13.9	18.0
Spain	€/MWh	9.8	12.5	13.4	17.3
Sweden	€/MWh	12.4	15.8	16.9	21.9
UK	€/MWh	10.2	13.0	13.9	18.0

Own calculations on the basis of data of the EU Commission, IEA, and EUROSTAT - [1]Peat

Table 7: Nominal price path of important primary energies in the EU, 2000 to 2020

---

[60] For details see RWE-Rheinbraun (2001), pp 12.

Besides brown coal hard coal as well covers a significant part of the energy demand in the EU. Production decreased in the last decade from nearly 180 Mtce to less than 90 Mtce. This drop was caused by the difference in production costs between countries outside of the EU and the European producers. The prices for hard coal of the EU member states are more and more based on the world market and less on the production costs. This leads to an equalisation of European hard coal prices. Differences are caused by the location of power plants, respectively by transport costs from the harbours to the inland locations.

Similarly the import prices for natural gas will converge. This process is significantly influenced by the competitive opening of national gas markets, which itself will stimulate the integration process of the European internal market. Additionally, the price path will be influenced by the increasing share of natural gas imports and the decrease of domestic production of EU member states. Remaining differences in the price level can be explained by differences in quality of natural gas depending on its origin and the transport from the production site to the location of the power plant, as well as by the specific demand characteristics of the single member states.

Besides the energy prices, the additional costs of an ET system will be influenced by several other factors: e.g. innovations and new techniques. In principle, Emissions Trading will set an incentive for the development of techniques which are so far not competitive or unknown. It is impossible to quantify these effects precisely. Therefore the following analysis will be restricted to a more conservative approach including only known techniques to forecast impacts on efficiency and substitution processes. The technical improvements described in the climate agreement of the German industry will be used as a basis to implement the possible range of emission reductions. This agreement implements nearly the same installations as the EU proposal.

6.3 Impact on the energy sector

As participants in an EU-wide ET system the EU proposal includes power plants, thermal power station, thermal plant, refineries, and coke plants, provided that their performance exceeds 20 MW. Therefore installations of the industry are concerned as well, even if other installations of the industrial production process will not be included in the ET system.[61] As power plants of the industry, mostly CHP, are optimised in a location-wide co-ordinated system, it seems useful to treat them separatedly from power plants of the energy sector. This chapter will focus on the power generation, specially on electricity.

6.3.1 Short-term effects

At least in the first period from 2005 to 2007, the ET is restricted on CO_2. Therefore fossil fuels will be affected only concerning their carbon content. Hard and brown coal will be specially affected in the pure power and heat generation and in CHP plants. In this context short-term and long-term effects have to be differentiated: Short-term effects base on unchanged power plants and can change the merit order of the given power plants by changing the relative prices of fossil fuels. The merit order is determined by the marginal costs of power generation, mainly the costs for fossil fuels (specific input of each fossil fuel times its price) and the price for the emission permits for combustion. How fast and in how far this change in relative prices will cause a substitution process depends on the initial allocation of permits. Principally, permits will be sold and the power generation will be reduced if the profit from selling permits exceeds the costs for fossil fuels. The production of electricity is expanded and the additionally needed permit is bought if the additional costs are lower than the additional profits of the higher electricity supply. The profits are affected only if at least a part of the initial allocation is sold and not given for free. Grandfathering has no impact on the costs of fossil fuels and therefore does not change the merit order. Even the assumed auctioning of 20% of the permits in the initial allocation will not change the merit order as the differences in marginal costs will not be affected, see table 8. The short-term reduction costs for all member states are lower than the prices for CO_2 in the year 2010 which resulted from chapter 5. This is also valid if the average efficiency of modern installations is compared with that of old ones. Additionally, legal aspects restrict the short-term substitution effects, like the remaining life expectancy defined in the large combustion plant directive of each member state. For instance, the Spanish brown coal power plants are fired with a mix of domestic (rich in sulphur) brown coal and imported (poor in sulphur) hard coal to keep the maximum SO_2 limit.

---

[61] This consequent division, which is likely to cause fundamental problems for the implementation of the EU proposal, was modified by the Danish compromise of August 2002 and, by that, come closer to real production conditions.

6.3.2 Long-term effects

The surprising result is, that the given power plants restrict the possibilities of short-term emission reductions by an ET system. Significant effects will be reached by an ET system, by the shut down of existing plants and the construction of new ones. It has to be kept in mind, that the investment decision regarding the construction of a new power plant does not only base on expected prices for emission permits but also on the usual costs for fossil fuel input, capital, employees, reparation and servicing, the input of factory and operating supplies and insurances.

The CO_2 reduction of a newly constructed power plant, the carbon content of the needed fossil fuel is as important as the efficiency of power generation. Normally the effectiveness is defined by the degree of efficiency, which is the relation between utilisable energy (measured in kilowatt hours) and the input of energy. Pure power generation plants which base on the combustion of fossil fuels and gain electricity by the steam turbine process, have a maximum in their degree of efficiency defined by Carnots Law. This technical and physical limit lies at 60% with the currently realisable steam parameters.

Fossil fuel	Brown coal			Hard coal		Heating oil
Substitute	Hard coal	Heating oil	Natural gas	Heating oil	Natural gas	Natural gas
Grandfathering						
Belgium				1003	149	241
Denmark				1121	119	192
Germany	163	434	91	1180	172	277
Finland	144	499	111	1357	208	336
Greece	206	521	78	1416	146	235
Italy				1121	176	284
Netherlands				1003	119	192
Portugal				1239	171	276
Spain				1180	176	283
UK				1121	142	228
Partial auctioning[1]						
Belgium				936	124	111
Denmark				1072	101	55
Germany	122	408	76	1124	151	133
Finland	117	482	100	1308	182	170
Greece	177	503	67	1345	124	62
Italy				1050	150	147
Netherlands				947	98	70
Portugal				1182	150	125
Spain				1123	154	139
UK				1052	115	91

Own calculations – [1]Auctioning of 20% of the permits of the base year 1999.

Table 8: Short term reduction costs in the electricity production, €/t CO_2 2010

A multitude of improvements and innovative elements caused an increase in efficiency of newly constructed power plants close to this physical limit. Hence, the gross degree of efficiency of hard coal fired plants constructed in the last decade was about 46%.[62] The gross degree of efficiency of brown coal

---

[62] The techncal parameter of hard coal fired plants constructed in the nineties is substantially identical with bloc 5 of the plant Staudinger, which started operation in the year 1993 by Preussen-Elektra. The degree of efficiency of this plant achieves a net degree of 42.5%, with an own con-

fired plants increased to nearly 45%.[63] As measured by the average degree of efficiency of the entire plant park (hard coal: 39.6%, brown coal: 35.3%) these progressions are noticeable. For the pure steam process this seems to be very close to the reachable maximum.

	2005	2010	2020
	in gce/kWh		
Nuclear energy	350	350	340
Brown coal, dry combustion	256	250	250
Brown coal, gas and steam co-generation plant		240	235
Hard coal, dry combustion	261	260	255
Hard coal, gas and steam cogeneration plant		255	245
Natural gas, gas and steam co-generation plant	320	315	310
Natural gas, gas turbine	215	210	205
	in g CO_2/kwh		
Brown coal, dry combustion	840	820	820
Brown coal, gas and steam co-generation plant		790	770
Hard coal, dry combustion	711	710	695
Hard coal, gas and steam cogeneration plant		695	670
Natural gas, gas and steam co-generation plant	525	517	510
Natural gas, gas turbine	350	345	335
Own calculations.			

Table 9: Specific fossil fuel input in power plants, 2005 to 2020

---

sumption of 8%. Therefore the gross degree is about 45.9%. For details see Rühland (1993), pp 624.

[63] The net degree of brown coal fired plants put on stream by the VEAG in the last years in East Germany are all above 40%. For details see VEAG (2000), p 22. The BOA plant put on stream in Septembre 2002 by RWE-Rheinbraun achieved a gross degree of efficiency stated at nearly 45%. For details see Jäger/Theis (2001), pp 21.

Significantly higher degrees of efficiency can be realised by gas and steam cogeneration plants. In principle, this process can be realised on the base of coal or natural gas. The input of coal currently entails substantial uncertainties in regard to the operating behaviour, the cleaning of sulphurous coal gases, and the part load behaviour. These problems were solved for natural gas fired gas and steam plants. Thus, for natural gas fired gas and steam plants a technically confirmed degree of efficiency of about 55%[64] can be assumed. It is unlikely that the technique for hard and brown coal fired gas and steam plants will be well-engineered before 2010, so that this type of plant will be taken into calculation for the period only after 2010. The potential increase in efficiency is specially very high in the use of brown coal due to its high water vapour content; yet, they will not achieve the degree of efficiency of natural gas fired gas and steam plants. The assumed increase in efficiency won't last to compensate the different carbon contents of the fossil fuels. Even in a perfectly optimised installation, the brown coal based electricity production emits twice as much CO_2 as a natural gas fired plant, see table 9.

Among fossil fuel costs the construction costs of new power plants significantly determine the costs of electricity production. Due to specific requirements of each fossil fuel on the installation, the differences in investment costs are noticeable. The construction of natural gas fired gas and steam cogeneration plants is comparatively competitive as no additional techniques for environmental protection (flue gas desulphurisation and removal) and no fossil fuel preparation is required. Investment costs for installed performance are about 500 €/kW, whereas the share of the gas turbine is around 400 €/kW, and the part of the steam turbine about 700 €/kW. Coal fired plants with a dry combustion technique cost twice as much. The construction of a nuclear power plant costs three times as much. The efficient techniques of hard and brown coal fired gas and steam fired plants are currently significantly more expensive than less efficient techniques, see table 10.

Considering financing costs in the construction period the construction costs arise from the specific installation costs. As the construction of a wind power plant lasts less than one year and that of a natural gas fired gas and steam cogeneration plant about two years, additional cost differences occur in comparison with hard and brown fired plants (construction time of 4 years) or with nuclear power plants (6 years). The specific construction costs are the basis for

---

[64] The gas and steam plant of Turbogas constructed by RWE-Power in Tapada, Portugal, achieves a net degree of efficiency of more than 55%. For details see w/o author (2000).

capital cost calculations. This calculation is deduced from dynamic investment account which implements interest rate and amortisations by instalments for a period of about 20 years.[65] As a nominal interest rate 6% is assumed, whereas no differentiation is made in regard to outside capital and own capital. The interest rate leads in connection with a duration of repayment of 20 years leads to an annuity of 8.72%.

	2005	2010	2020
Nuclear energy	1 740	1 920	2 180
Brown coal, dry combustion	1 125	1 240	1 410
Brown coal, gas and steam cogeneration plant		1 470	1 670
Hard coal, dry combustion	1 020	1 130	1 285
Hard coal, gas and steam cogeneration plant		1 470	1 670
Natural gas, gas and steam cogeneration plant	410	450	515
Natural gas, gas turbine	510	565	640
Own calculations.			

Table 10: Specific investment costs of new power plant constructions, 2005 to 2020 in €/kW

Personnel costs, costs for repair and servicing, costs for operating supplies, and costs for liability insurance, are combined as the rest of costs. In comparison with the costs for fossil fuels and capital costs they matter less in an investment calculation. They are neither specified subjects to the installed performance nor to the hours worked. Personnel costs, costs for repair and servicing, and insurance costs can be rated as production-based. Personnel costs depend on the number of employees in each type of power plant, on their average annual working time, their nominal wage, and the statutory and voluntary social insurance contribution. The average working time, the wages and the social insurance contribution will not be differentiated for each type of power plant in the following analysis. Therefore, differences in personnel costs result solely from the number of needed employees in each type of power plant. Nuclear

---

[65] A substance conservation is not integrated in this method of depreciation, as the investments will be valuated for purchase prices and not for replacement costs.

power plants require 350 persons, hard coal fired plants 220, brown coal fired plants about 240, and natural gas fired plants 80 persons.

The sum of all costs and the unification in reference to the block size of 1 300 Megawatt (MW) and its accumulation for a period of 20 years leads to the conclusion that without the introduction of an ET system brown coal will remain the most competitive fossil fuel until 2020 in power generation. Brown coal fired plants can produce electricity much more competitivly than any other fossil fuel, including natural gas fired gas and steam cogeneration plants, see table 11. In particular, the construction of new nuclear power plants cannot compete with brown coal fired plants. This is also valid in comparison with new hard coal fired plants. The difference between these types is quite high with 10 €/MWh respectively 7 €/MWh. This cost difference cannot be compensated by an increasing efficiency. Natural gas or coal fired gas and steam plants are not the cheapest technique in the considered period. Irrespective of CO_2 emission restrictions, on the basis of traditional profitability calculations gas and steam cogeneration plants will not come into operation for the base load within the next 20 years.

	2005	2010	2020
	Base load		
Nuclear energy	49.8	55.5	63.6
Brown coal, dry combustion	38.4	42.7	50.5
Brown coal, gas and steam co-generation plant		48.6	56.5
Hard coal, dry combustion	40.3	44.6	52.6
Hard coal, gas and steam cogeneration plant		51.0	59.8
Natural gas, gas and steam co-generation plant	49.9	54.2	61.9
	Middle load		
Nuclear energy	79.9	89.5	103.8
Brown coal, dry combustion	58.7	64.5	77.4
Brown coal, gas and steam co-generation plant		75.7	88.6
Hard coal, dry combustion	58.1	64.6	76.3
Hard coal, gas and steam cogeneration plant		77.1	90.6
Natural gas, gas and steam co-generation plant	59.5	64.9	74.4
Own calculations.			

Table 11: Long-term production costs of new constructed power plants, 2005 to 2020, in €/kWh

Currently the middle load range is dominated by hard coal. This will remain the same for the next 10 to 20 years, if energy and environmental conditions stay unchanged. Though the construction of brown coal fired plants is profitable, the availability of brown coal and its technical restrictions within the discontinuous operation mode of a middle load power plant contradict its input for this range. The cost disadvantages of hard and brown coal fired gas and stream plants are worse for the middle load range in comparison with the base load range because the higher-than-average capital costs have to be spread on a smaller production cost. This disadvantage does not apply to natural gas fired gas and steam technique due to its very low capital costs. Thus, this type of power plant already improves its competitive position towards hard coal fired plants in the middle load range without the introduction of an ET system for CO_2. At the beginning of the analysis period, the production costs are even, but then change continuously for the benefit of natural gas. The low cost difference proves of the possibility of significant impacts of only small changes within the general conditions.

The installation of an ET system causes such a significant change in general conditions. Within profitability calculations the value of emission permits has to be implemented additionally. Particularly, the investment decision about the construction of a new power plant includes the possibility to sell uncommitted permits. The decision between the use of permits for a specific technique respectively a specific fossil fuel and the disposal on the market depends solely on the permit price. Particularly notable is the fact that this decision is absolutely detached from the initial allocation. Grandfathering does not engage an operator of a brown coal fired power plant to use the permits only for the use of this type of installation.

Fossil fuel	Substitute	2005	2010	2020
Base load				
Brown coal	Nuclear energy	13.5	15.2	16.0
	Hard coal	14.4	14.6	17.1
	Natural gas fired gas and steam plant	23.5	23.1	23.6
Hard coal	Nuclear energy	13.3	15.3	15.8
	Natural gas fired gas and steam plant	26.8	26.1	25.9
Middle load				
Brown coal	Nuclear energy	25.2	28.6	32.3
	Hard coal	- 4.9	- 6.8	- 8.7
	Natural gas fired gas and steam plant	1.6	- 1.2	- 6.1
Hard coal	Nuclear energy	30.6	35.1	39.7
	Natural gas fired gas and steam plant	3.9	0.8	- 5.2
Own calculations.				

Table 12: Long-term marginal reduction costs in power generation, 2005 to 2020, in €/t CO_2

Due to these realities the cost disadvantage of nuclear energy turns to a significant advantage: already for a CO_2 price of about 15 €/t, power generation for new plants of this type will be more competitive than those of brown or hard coal fired plants, see table 12. This benefit rises with an increasing carbon price. Within the EU proposal not only the reinvestments in nuclear power plants would be seen as profitable, but also the construction of new plants of this type as a replacement of brown and hard coal fired plants. Gas and steam cogeneration plants can also improve their competitive position according to their higher efficiency. Admittedly, the efficiency based benefits will be more than counterbalanced by the difference in specific carbon contents. On balance, the EU proposal will make the construction of natural gas fired gas and steam cogeneration plants for the base load profitable, even for low carbon prices. Thus, in the long-term hard and brown coal fired plants will be displaced by newly constructed natural gas fired plants.

	2000 to 2010	2010 to 2020	2000 to 2020
	Brown coal		
Germany	3.6	6.0	9.6
Greece	0.7	1.1	1.8
Finland[2]	0.8	0.4	1.2
Ireland[2]	0.4	-	0.4
Spain	1.1	-	1.1
EU 15 total	6.6	7.5	14.1
	Hard coal		
Belgium	1.6	0.7	2.3
Denmark	0.7	3.4	4.1
Germany	8.4	8.2	16.6
Finland	1.1	1.5	2.6
France	9.1	3.8	12.9
Italy	0.5	1.7	2.2
Netherlands	1.2	-	1.2
Spain	2.7	1.0	3.7
UK	16.3	15.6	31.9
EU 15 total	41.6	35.9	77.5

Own calculations and data of the EU commission – [1]for an average useful life of 40 years; [2]Peat.

Table 13: Long-term substitution potentials in power generation[1], 2010 and 2020, in GW

In an ET scheme with the possibility to assign reductions from JI and CDM projects these serious shiftings will not occur within the next 20 years. The base load structure will widely remain with a CO_2 price of 5 €/t. The displacement of hard coal by natural gas in the middle load range will be indistinct. Incentives for a fuel switch set by the initial allocation are very low for a price of 5 €/t CO_2.

	Brown coal	Hard coal	Total
		2010	
Belgium		-4.0	-4.0
Denmark		-1.3	-1.3
Germany	-18.2	-18.4	-36.6
Finland	-2.6	-2.0	-4.3
France		-12.5	-12.5
Greece	-3.5	0	-3.5
Ireland	-1.6	0	-1.6
Italy		-1.5	-1.5
Netherlands		-3.6	-3.6
Spain	-5.0	-6.8	-11.8
UK		-29.8	-29.8
EU 15 total	-30.6	-79.8	-110.4
		2020	
Belgium		-5.8	-5.8
Denmark		-7.5	-7.5
Germany	-48.5	-36.4	-84.9
Finland	-3.5	-4.7	-8.2
France		-17.7	-17.7
Greece	-9.1	0	-9.1
Ireland	-1.6	0	-1.6
Italy		-6.6	-6.6
Netherlands		-3.6	-3.6
Spain	-5.0	-9.3	-14.3
UK		-58.2	-58.2
EU 15 total	-67.7	-149.7	-217.4

Own calculations.

Table 14: Substitutional CO_2 reductions in power generation, 2010 to 2020, Mt CO_2

Of particular importance for the following analysis is the factor that the EU proposal will not affect all member states similarly. The impact will be focused on a few member states. Particularly those locations will be affected, which reach their economic life-time within the next 20 years and which have to be replaced by new installations.

Regarding brown coal, this affects mainly Germany, more precisely, the left-Rhine area. Regarding hard coal Germany is affected as well and also France and specially UK. Insofar it is not surprising, that most of the CO_2 reduction can be expected from these member states. Until 2012 an amount of 110 Mt CO_2 can be reduced by replacing old hard and brown coal fired plants by natural gas fired gas and steam cogeneration plants. One third of this amount would account for the German power generation sector. An additional modernisation in the decade after 2012 can lead to similar reductions. Therefore a fuel switch can reduce CO_2 emissions in the EU until 2020 by 217 Mt, see table 14. Admittedly, this fuel switch strategy will not be sufficient to cover the Kyoto commitment.

The displacement of hard and brown coal by natural gas causes a fundamental change in the energy and cost structures of power generation. Power plants with low capital costs take over the place of those with high capital costs. Thereby the relevance of fossil fuel costs rises significantly and the dependency on fossil fuel imports increases. This automatically implements an increase in price and supply risks, which are, as experience has shown, higher for natural gas than for hard coal imports. Table 15 summarizes the cost impulses. They depend on the assumption that the additional quantities (until 2020 around 545 TWh) of natural gas imports will be supplied for the same price. Price increases will not question the fuel switch but will be able to cause much higher additional costs for the production of electricity.

	Fossil fuel	Other costs[1]	Total
		2010	
Belgium	131.4	-94.7	36.7
Denmark	50.0	-41.4	8.6
Germany	1 512.7	-744.4	768.3
Finland	240.0	-120.1	119.9
France	579.0	-538.4	40.6
Greece	108.4	-48.1	60.3
Ireland	58.3	-27.5	30.8
Italy	64.2	-29.6	34.6
Netherlands	125.5	-71.0	54.5
Spain	343.9	-159.7	184.2
UK	819.0	-964.3	-145.3
EU 15 total	4 032.4	-2 839.1	1 193.3
		2020	
Belgium	246.4	-136.1	110.3
Denmark	380.8	-242.6	138.3
Germany	4 419.6	-1 729.6	2 690.0
Finland	605.2	-247.3	357.9
France	1 067.7	-763.2	304.6
Greece	361.2	-140.2	221.0
Ireland	76.3	-31.1	45.1
Italy	323.1	-130.2	192.9
Netherlands	163.4	-71.0	92.5
Spain	549.2	-218.9	330.3
UK	2 095.3	-1 887.2	208.1
EU 15 total	10 288.3	-5 597.1	4 691.1

Own calculations – [1]Costs for capital and personnel, and the rest of costs.

Table 15: Substitutional cost impulses in power generation, 2010 to 2020, Mio. €

Independent of these changes in the energy mix the charging of fossil fuels concerning their carbon content will cause fundamental cost distortions between the power generation of each member state. Even though the structural change in the power plant park is massive, the additional costs cannot be neutralised even if the profit from selling permits is considered, the predominant part will be cost and price effective. This cost impulse depends on the CO_2 price, specially on the kind of initial allocation. An auctioning of 20% of the initial allocation will lead to additional costs twice as high as the proper emission reductions, which are caused by the switch from hard and brown coal to natural gas.

	ET incl. JI / CDM	EU proposal		
	Partial auctioning[1]	Reduction costs	Partial auctioning[1]	Total
2010				
Belgium	0.30	0.45	1.81	2.27
Denmark	0.74	0.23	4.46	4.69
Germany	0.63	1.49	3.78	5.27
Finland	0.31	1.80	1.83	3.63
France	0.08	0.08	0.51	0.59
Greece	0.86	1.32	5.13	6.45
Ireland	0.76	1.47	4.55	6.03
Italy	0.54	0.14	3.22	3.36
Netherlands	0.67	0.66	4.04	4.70
Austria	0.22	0.00	1.35	1.35
Portugal	0.58	0.00	3.50	3.50
Sweden	0.04	0.00	0.27	0.27
Spain	0.47	0.92	2.80	3.72
UK	0.47	-0.42	2.81	2.40
EU 15 total	0.41	0.50	2.47	2.96
2020				
Belgium	0.23	1.38	1.36	2.75
Denmark	0.54	3.25	3.74	6.98
Germany	0.47	2.80	5.20	8.00
Finland	0.18	1.09	5.37	6.46
France	0.05	0.29	0.61	0.90
Greece	0.66	3.94	4.83	8.76
Ireland	0.68	4.09	2.16	6.25
Italy	0.51	3.07	0.76	3.83
Netherlands	0.63	3.78	1.11	4.89
Austria	0.22	1.35	0.00	1.35
Portugal	0.58	3.50	0.00	3.50
Sweden	0.04	0.27	0.00	0.27
Spain	0.40	2.37	1.66	4.03
UK	0.30	1.81	0.60	2.41
EU 15 total	0.32	1.92	1.95	3.87

Own calculations – [1]20 % of the emissions in the year 1999

Table 16: ET and costs of power generation, 2010 to 2020, €/MWh

Member states and regions with a high share of carbonaceous energy sources in their whole energy input will suffer from above-average additional expenses. Those with a high share of nuclear energy or regenerative energy sources will get just a marginal burden, see table 16. In comparison with, for instance, Sweden, whose power generation is almost completely covered by hydropower and nuclear energy, the EU proposal will induce in Germany or member states with comparable fossil fuel input additional costs that will be more than ten times higher.

6.3.3 Combined heat and power cycle

Combined heat and power cycles get a significant meaning in this discussion. The reduction potential until 2005 is estimated with 10 Mt, until 2010 with 23 Mt CO_2. Decisive for this reduction are technical modifications in the coupling of power and heat generation, specially the displacement of pure steam turbines by gas and steam techniques. The coupling of gas and steam turbines allows a doubling in power generation while the heat generation remains constant. Thereby, the increase in power generation is caused only to a small degree by the efficiency improvement, and much more by the boost in installed performance. As mentioned above, the gas and steam technique is independent from the choice of fossil fuels but is currently only realised for natural gas.

Although the meaning of combined heat and power cycles is undisputed the adequate instrumentation is controversial. Currently, positive experiences lead to a preference of self-commitments, as agreed between the German government and the German industry. This agreement defines a CO_2 reduction compared with 1998 of 10 Mt until 2005, and of 20 to 23 Mt until 2010, by the conservation, the modernisation and the new construction of combined heat and power cycles (CHP).

These target shall be reached by a bonus system which is defined in the CHP law. Therein, the bonus system will only apply to existing installations differentiated by age and modernity. It will be limited in time and designed digressively, and orientate itself at the quantity of energy production. The promotion of installations put on stream before 01.01.1990 is limited at a maximum of 5 years. Newer CHP installations can be promoted for 8 years. Longer promotions will only be given for installation whose investment exceeds 50% of a new construction and which will be modernised fundamentally.

The introduction of an ET scheme will have a significant impact on this carbon abatement measure. Similar to the electricity sector not only the short-term marginal costs have to be considered but also the long-term marginal costs. In this comparison, the additional factor that a modernisation of a, for instance, hard coal fired steam turbine increases efficiency but not the process itself has to be considered. This specially applies to the quantity of electricity production. In contrast, the transition from a steam turbine technique to a gas and steam cogenerated technique allows a significantly higher production of electricity.[66]

For new constructions natural gas fired gas and steam cogenerated plants are more competitive in all load ranges than hard and brown coal fired ones, see table 17. Decisive for this comparison are the investment costs which are half as much for the natural gas fired type as for the coal fired type of gas and steam cogenerated plant. The low capital costs lead to a disadvantage for hard and brown coal, specially for a duration of use of less than 4,000 hours per annum. For 2,000 hours per annum, which is typical for the current use of CHP plants, a hard or brown coal fired cogeneration plant causes nearly twice as much costs as natural fired gas and steam cogeneration plants. Therefore, natural gas will gain an increasing significance in the public district heat supply.

The cost relation shifts for the benefit of hard and brown coal in the case of a modernisation of old installations as, for a time span of at least 8 years, the electricity produced in CHP plants will be paid at 0.15 c/kWh. Additionally, the investment costs are much lower than for new constructions even if the modernisation costs exceed at least 50% of the investment of a new construction (bonus arrangement). Thereby the specific cost advantages of a gas and steam cogeneration plant can be compensated. These competition improvements confront unaltered the revenues of a new gas and steam plant which are twice as high. This aspect will only then not lead to new constructions of natural gas fired gas and steam plants, if the feeding-in fees are less than 1.7 c/kWh. This is not likely even in the case of a liberalisation. Insofar the construction of a new natural gas fired gas and steam cogeneration plant is more competitive than a modernisation of an existing plant on the base of hard or

---

[66] For the following calculations it is assumed that a steam turbine with a combustion heat performance of 100 MW has an overall efficiency of 90% and an electricity index of 0.45. Therefore the maximal heat extraction is 45MW. In a gas and steam cogeneration plant with an identical heat extraction and the same overall efficiency the combustion heat performance is nearly 170 MW. Here it is assumed that the efficiency of the gas turbine is 38%. The electrical power of a gas and steam turbine sums up to 83 MW with an electricity index of 1.85. Additional costs have been taken over by the comparison of the power generation costs, whereas the specific numbers were fit to the installation size of 100 MW.

brown coal, even if the bonus arrangement is considered. This applies specially in an ET scheme which additionally burdens the carbon content of fossil fuels.

Fossil fuel	2005	2010	2020
	Base load		
Brown coal	38.8	43.5	49.4
Hard coal	36.3	40.5	46.6
Natural gas	30.4	33.7	38.7
	Middle load		
Brown coal	63.4	71.7	80.7
Hard coal	57.4	64.2	73.7
Natural gas	37.5	73.7	47.7
	Spitzenlast		
Brown coal	120.7	135.7	153.7
Hard coal	106.5	119.5	137.0
Natural gas	54.1	60.3	68.7
Own calculations.			

Table 17: Long-term production costs in new constructed Combined heat power cycles, 2005 to 2020, €/MWh

6.4 Sectoral impacts

As already mentioned at the beginning, specific application conditions apply to industrial power generation. These conditions differ significantly from the general power supply. Industrial installations are predominantly used as CHP plants, which supply the whole location with electricity and heat. Therefore the emission reduction costs are comparable with the costs for CHP plants analysed in the previous paragraph. There, one conclusion was that the replacement of existing hard or brown coal fired plants by natural gas fired gas and steam cogeneration plants is already competitive without additional incentives. This is reflected by the modernisations of the last years. Increasingly, these modernisations are realised by so-called contracting models. These types of models make a distinction between the construction and the line operation of a new installation from the use of the produced quantity of power and heat. Responsible for the construction and the line operation is a public supply company and not the industry itself. But the installation is designed in regard to specific needs of the industry company. The CHP co-plant for BASF in Ludwigshafen was realised in this custom-made manner: Owner and operator of the plant is RWE Power AG, performance and installation configuration are designed for the demand of BASF for about 6 TWh electric energy and about 18 Mt process steam.

For the future one can expect more of such contracting projects. The monitoring report of the VDEW lists the realised and planned projects for the year 1998. Accordingly, 17 GHP co-operation project were realised between 1996 and 1999 and 7 more projects are in planning. The electrical power of these projects sum up to a total of about 2 500 Megawatt. The largest projects are the GHP cogeneration plant of Bayer AG in Dormagen (450 MW), the GHP cogeneration plant of the Adam Opel AG in Rüsselsheim (100 MW), the GHP cogeneration plant in Eisenhüttenstadt (110 MW), GHP cogeneration plant of the Henkel AG (300 MW), and the GHP cogeneration plant of the Hoechst AG in Frankfurt (about 500 MW). While the date of realisation stays open for the two projects mentioned last, the other projects are already realised or in concrete planning.

These new installations contribute significantly to the reduction of CO_2 emissions. If one assumes for each plant park similar reductions as have been realised in Ludwigshafen by the construction of the CHP plant of the RWE Energie AG, the CO_2 emissions could be reduced by 8 to 10 Mt. A part of these reductions was already realised until 1998. Yet, the bigger part will be realised within the next years and contribute to the achievement of a reduction of CO_2 emissions of 28% by 2005 and of 35% until 2012, agreed upon in the climate agreement of the year 2000.

Other member states show similar developments. Emission reductions caused by this plant modernisation process will be realised independently from the introduction of an ET scheme for CO_2. Therefore, additional reductions can only be caused by efficiency improvements and a further fuel switch within the real industrial processes. In this area, the technical state-of-the-art has comprehensively improved as well, so that an additional increase in the energy efficiency can only be realised with very high additional costs. Monitoring reports document the increase in efficiency of industrial production installations in Germany since 1996.[67] These reports document a range of emission reduction

---

[67] Cf. Buttermann/Hillebrand (2002a).

measures from the modernisation of existing installations, the construction of new, and the shutdown of old installations, up to newly developed concepts of integrated supply. Thereby it has to be stated, that specially in the industrial production of basic materials, the efficiency improvements become more and more cost intensive and even hit scientific technical limits, which exclude additional improvements.

	Steel	Stone, Ceramic, glass	Pulp/paper	Chemistry	Non-iron metals
Belgium	2.36	0.43	0.08	0.25	0.10
Denmark	0.12	0.57	0.12	0.06	0.05
Germany	1.58	0.53	0.26	0.27	0.18
Finland	1.92	0.85	0.27	0.37	0.09
France	1.41	0.58	0.26	0.32	0.26
Greece	0.29	3.38	0.61	0.38	1.10
Ireland	0.22	0.65	0.06	0.11	0.00
Italy	1.08	1.11	0.29	0.47	0.25
Netherlands	4.58	0.41	0.19	0.65	0.10
Austria	1.85	0.34	0.24	0.24	0.11
Portugal	0.74	1.16	0.53	0.38	0.28
Sweden	0.95	0.61	0.14	0.19	0.22
Spain	0.92	0.83	0.32	0.29	0.25
UK	2.01	0.49	0.28	0.69	0.36
Own calculations.					

Table 18: Specific CO_2 emissions for each sector and member state 1999, t CO_2/1000 € gross production

Excursion

Table 18 refers to significant statistical problems for the survey and quantification of specific CO_2 emissions. As the Dutch steel industry is similarly efficient as the German one, it must be for the denominator (gross production) that the values differ that much. As table 18 includes the currently best available data base, fundamental problems in connection with the realisation of an initial allocation neutral to competition are indicated.

Additional reduction potentials to basic efficiency improvements can be opened up by the fuel switch from brown and hard coal to natural gas. Admittedly, it has to be considered, that the consumption structures are mostly caused by the technical and physical legality of each production process. Thus, the above-average specific CO_2 emissions of the Belgian and Dutch steel production can be explained by the dominant role of oxygen steel, see table 18.[68] Emission reductions are in these cases mostly tantamount to efficiency improvements or fundamental changes in production processes. Therefore reductions do not only depend on the price for carbon but also on other determinants like product quality, availability of raw materials, or other process costs. As coke is needed as a rack to gas the Moeller's column, coal and coke input for blast furnaces cannot be replaced by natural gas. Additionally, by a higher natural gas input the temperature profile of a furnace is changed, so that more oxygen has to be put in the recuperator, which requires higher investments.[69]

	2005	2010	2020
	€/t CO_2		
Germany	123	131	173
France	116	123	163
Greece	92	98	131
Italy	98	105	139
Portugal	126	135	178
Sweden	147	160	213
Spain	111	120	157
UK	103	107	147
	€/t clinker		
Germany	15	16	21
France	14	15	20
Greece	11	12	16
Italy	12	13	17
Portugal	15	17	23
Sweden	18	19	26
Spain	14	15	19
UK	13	13	18
Own calculations.			

Table 19: Long-term reduction costs in the cement production, 2005 to 2020

Even if a substitution of hard and brown coal by natural gas is technically possible, for instance by replacement investment, the additional costs are signifi-

---

[68] It is important for the interpretation of the coefficients that the CO_2 emissions do not refer to physical production but to the real gross production value. Therefore they differ from those ones used in technical studies.

[69] For details see Peters/Schmölde (2002), pp 43.

cantly high. Exemplary, table 19 displays the cost effects for the cement production of the EU member states. These calculations make significant cost differences between the member states clear. On the other hand, the deterioration of the competitiveness of all EU member states towards third countries is noticeable. To remove additional costs of 20 €/t clinker at the market, the cement prices have to be lifted by 50%. This transmittance seems to be impossible regarding the competitors from third countries. If the cement industry was obliged to a fuel switch by the restrictive concept of an ET scheme, relocations would be forced. This effect will not occur until 2010 as the cement industry does not have to implement its own reductions at a price of 30 €/t CO_2 but, instead, will have to buy the needed 20% of emission permits in the partial auctioning.

As similar reactions can be expected in other industry sectors, the focus of the following analysis lies on the sectoral additional costs, which are caused by the purchase of permits in the partial auctioning or by the indirect cost effects of the electricity price rise. Cost and price effects will be displayed for those sectors which are included in the official enquiry. This data base gives a consistent structuring and complete data for all member states, specially for the energy demand, production, and exports and imports.

The sectors are mainly

- the chemical industry (WZ-No 24),
- the sector glass, ceramics, and stones and earth (WZ-No 26),
- the iron producing industry (WZ-No 27 1 bis 27 3)
- the non-iron metal production and processing (WZ-No 27 4), and
- the sector pulp and paper (WZ-No 21).

"CO₂ Emissions Trading put to test
– Design problems of the EU draft directive concerning an Emissions Trading scheme"

	ET incl. JI and CDM		EU proposal	
	Electricity costs	Total	Electricity costs	Total
	2010			
Belgium	1.9	20.6	14.0	126.6
Denmark	0.5	0.6	2.9	3.6
Germany	13.9	75.3	116.4	484.7
Finland	0.8	8.1	9.2	53.0
France	1.3	31.1	9.0	187.6
Greece	0.7	1.0	5.3	7.0
Ireland	0.2	0.3	1.9	2.3
Italy	9.9	32.0	61.7	194.5
Netherlands	1.5	15.4	10.6	93.9
Austria	0.5	8.8	2.9	52.9
Portugal	0.5	1.4	2.7	8.2
Sweden	0.2	6.7	1.3	40.2
Spain	6.2	16.9	49.4	113.9
UK	4.6	31.3	23.7	183.6
EU 15 total	42.6	249.5	311.1	1552.0
	2020			
Belgium	1.4	20.2	17.0	129.6
Denmark	0.3	0.4	4.2	5.0
Germany	10.4	71.8	176.8	545.0
Finland	0.5	7.8	16.4	60.2
France	0.8	30.5	13.8	192.4
Greece	0.5	0.8	7.2	8.9
Ireland	0.2	0.3	1.9	2.4
Italy	9.4	31.5	70.3	203.1
Netherlands	1.4	15.3	11.1	94.4
Austria	0.5	8.8	2.9	52.9
Portugal	0.5	1.4	2.7	8.2
Sweden	0.2	6.7	1.3	40.2
Spain	5.3	16.1	53.5	117.9
UK	3.0	29.6	23.9	183.8
EU 15 total	34.3	241.1	403.1	1644.0
Own calculations.				

Table 20: ET and additional costs in the steel production, 2010 to 2020, Mio. €

"CO$_2$ Emissions Trading put to test – Design problems of the EU draft directive concerning an Emissions Trading scheme"

	ET incl. JI and CDM		EU proposal	
	Electricity costs	Total	Electricity costs	Total
	2010			
Belgium	0.7	3.7	5.6	23.2
Denmark	0.6	1.9	3.9	11.8
Germany	9.4	28.2	78.2	191.1
Finland	0.3	1.7	3.1	11.6
France	0.8	11.0	5.7	66.6
Greece	1.7	5.5	12.8	35.6
Ireland	0.5	1.3	3.6	8.8
Italy	7.0	27.5	43.8	166.7
Netherlands	1.0	2.8	7.3	18.1
Austria	0.3	1.7	1.8	10.4
Portugal	1.3	6.5	7.7	39.0
Sweden	0.1	1.2	0.3	7.3
Spain	4.1	17.4	32.7	112.6
UK	3.4	8.8	17.3	49.6
EU 15 total	31.1	119.2	223.7	752.4
	2020			
Belgium	0.6	3.5	6.8	24.4
Denmark	0.5	1.8	5.9	13.7
Germany	7.0	25.8	118.7	231.6
Finland	0.2	1.6	5.5	14.0
France	0.5	10.6	8.7	69.6
Greece	1.3	5.1	17.3	40.2
Ireland	0.4	1.3	3.7	8.9
Italy	6.6	27.1	49.9	172.9
Netherlands	1.0	2.8	7.6	18.4
Austria	0.3	1.7	1.8	10.4
Portugal	1.3	6.5	7.7	39.0
Sweden	0.0	1.2	0.3	7.3
Spain	3.5	16.8	35.4	115.3
UK	2.2	7.5	17.4	49.7
EU 15 total	25.3	113.4	286.7	815.4
Own calculations.				

Table 21: ET and additional costs of the sector glass, ceramics, and stones and earths, 2010 to 2020, Mio. €

Though, this sectoral level displays a broad picture of the economic impacts, the charges and discharges cannot be taken as representative for all aggregated subsectors. This specially applies for those sectors, which aggregate energy intensive basic material productions and labour intensive processings. The product range of the chemical industry, for instance, includes the production of energy intensive organic and inorganic basic materials, the production of important intermediates like varnish and paint, fertilisers, plant protective agents, or synthetics in primary forms, and the production of consumption near end products like drugs or detergents, which is produced with a relatively low specific energy input. A more disaggregated analysis, for instance for each production process cannot be made due to the lack of available EU-wide data. Insofar, the following calculations can only be interpreted as average values, which count for the sector as a total and from which single subsectors can differ significantly. Obviously, the burdens depend on the structure of the energy input. The burden is very high for sectors with an above-average input of carbon-rich fuel input. In this context the steel industry, see table 20, and the production of the sector glass, cement, lime, and stones and earths, see table 21, have to be listed.

An additional problem for the production of cement, lime, and glass, is the fact that the CO_2 emissions are caused by chemical reactions within the production process, for instance in the division of lime from limestone. Therefore, these emissions cannot be lowered by efficiency improvements or by a fuel switch. The inclusion of this type of CO_2 emissions in an ET scheme increases the burden of this sector significantly. The quantification of thus caused additional costs for each member state is very difficult: even the CO_2 emissions from the combustion of fossil fuels is unknown for this aggregation level, just as little, as the production quantities which are needed for a quantification of raw material caused CO_2 emissions. Solely the additional costs in the cement production can be approximated. But only with the restriction that a fix proportion between cement and clinker exists for all EU member states, which correspond with that known in Germany. With this restriction the additional costs (displayed in table 22) result. It can be seen, that a predominant part of the additional costs are caused by the purchase of permits to cover the emissions caused by raw material input and that the permits will cost six times as much in the EU proposal as in an ET scheme with JI and CDM.

	ET incl. JI and CDM		EU proposal	
	Energy	Raw material	Energy	Raw material
Belgium	2.7	4.0	16.9	24.3
Germany	11.5	15.1	74.6	90.8
France	5.4	8.7	32.8	52.1
Greece	5.4	6.7	34.8	40.0
Italy	12.4	16.8	74.7	100.6
Netherlands	1.2	1.5	7.2	9.0
Austria	1.1	1.6	6.5	9.8
Portugal	3.3	4.4	19.9	26.6
Sweden	0.7	1.2	4.3	7.0
Spain	11.8	16.4	74.7	98.6
UK	3.9	5.5	23.0	32.8
Own calculations.				

Table 22: ET and estimated additional costs of the cement production, 2010, Mio. €

	ET incl. JI and CDM		EU proposal	
	Electricity costs	Total	Electricity costs	Total
2010				
Belgium	0.6	1.0	4.3	6.7
Denmark	0.1	0.1	0.4	0.5
Germany	11.5	14.3	95.7	112.9
Finland	0.6	0.7	7.4	8.1
France	0.8	2.6	5.5	16.6
Greece	3.0	4.2	22.7	30.1
Ireland	0.3	1.1	2.0	7.0
Italy	2.8	4.2	17.6	25.8
Netherlands	3.8	4.0	26.7	27.9
Austria	0.1	0.3	0.5	1.7
Portugal	0.1	0.1	0.4	0.7
Sweden	0.1	0.4	0.7	2.3
Spain	4.2	5.2	33.5	39.2
UK	2.8	4.9	14.1	26.8
EU 15 total	30.7	43.1	231.6	306.4
2020				
Belgium	0.4	0.8	5.2	7.7
Denmark	0.0	0.1	0.6	0.7
Germany	8.5	11.4	145.3	162.5
Finland	0.4	0.5	13.1	13.8
France	0.5	2.3	8.5	19.5
Greece	2.3	3.6	30.8	38.2
Ireland	0.2	1.1	2.1	7.1
Italy	2.7	4.0	20.0	28.3
Netherlands	3.6	3.8	27.8	29.0
Austria	0.1	0.3	0.5	1.7
Portugal	0.1	0.1	0.4	0.7
Sweden	0.1	0.4	0.7	2.3
Spain	3.6	4.6	36.3	42.0
UK	1.8	3.9	14.2	26.9
EU 15 total	24.3	36.8	305.7	380.5
Own calculations.				

Table 23: ET and additional costs of the Non-iron metal production , 2010 and 2020, Mio. €

	ET incl. JI and CDM		EU proposal	
	Electricity costs	Total	Electricity costs	Total
2010				
Belgium	0.7	1.1	5.6	8.0
Denmark	0.5	0.7	3.4	4.3
Germany	12.6	19.6	105.3	147.2
Finland	7.5	11.7	89.4	114.5
France	1.0	5.9	6.9	36.3
Greece	0.4	0.7	3.1	5.1
Ireland	0.1	0.2	1.0	1.2
Italy	5.2	9.4	32.4	57.7
Netherlands	2.5	3.3	17.2	22.4
Austria	1.0	2.1	5.8	12.6
Portugal	1.1	2.3	6.9	13.6
Sweden	1.0	2.8	5.9	16.8
Spain	2.6	5.5	20.4	38.1
UK	5.1	8.6	26.1	46.8
EU 15 total	41.3	73.8	329.4	524.4
2020				
Belgium	0.6	1.0	6.8	9.2
Denmark	0.4	0.5	5.0	5.9
Germany	9.4	16.4	159.8	201.7
Finland	4.4	8.6	159.2	184.2
France	0.6	5.5	10.6	40.0
Greece	0.3	0.6	4.2	6.2
Ireland	0.1	0.1	1.0	1.2
Italy	4.9	9.1	37.0	62.2
Netherlands	2.3	3.2	18.0	23.1
Austria	0.9	2.1	5.8	12.6
Portugal	1.1	2.3	6.9	13.6
Sweden	0.9	2.7	5.9	16.8
Spain	2.2	5.1	22.1	39.7
UK	3.3	6.7	26.3	47.0
EU 15 total	31.5	64.0	468.5	663.5
Own calculations.				

Table 24: ET and additional costs of the sector pulp and paper, 2010 and 2020, Mio. €

	ET incl. JI and CDM		EU proposal	
	Electricity costs	Total	Electricity costs	Total
2010				
Belgium	3.9	10.9	29.0	71.5
Denmark	0.9	1.2	5.4	7.4
Germany	32.2	65.6	269.5	469.5
Finland	1.3	3.0	15.9	25.8
France	2.1	27.4	14.7	166.3
Greece	1.0	1.9	7.9	13.1
Ireland	0.7	2.4	5.8	15.9
Italy	11.2	32.3	70.1	196.8
Netherlands	8.0	25.5	55.7	161.1
Austria	0.5	1.8	2.9	10.8
Portugal	1.2	2.7	7.5	16.0
Sweden	0.3	2.0	1.6	12.1
Spain	5.0	12.9	39.8	87.5
UK	9.3	43.0	47.6	249.6
EU 15 total	77.7	232.7	573.3	1 503.4
2020				
Belgium	2.9	10.0	35.2	77.7
Denmark	0.6	1.0	8.0	10.1
Germany	24.0	57.4	409.2	609.2
Finland	0.8	2.4	28.3	38.2
France	1.3	26.5	22.6	174.1
Greece	0.8	1.7	10.7	15.9
Ireland	0.7	2.3	6.1	16.1
Italy	10.6	31.8	79.9	206.6
Netherlands	7.5	25.0	58.0	163.4
Austria	0.5	1.8	2.9	10.8
Portugal	1.2	2.7	7.5	16.0
Sweden	0.2	2.0	1.6	12.1
Spain	4.3	12.2	43.0	90.8
UK	6.0	39.6	47.8	249.9
EU 15 total	61.4	216.4	760.8	1 690.9

Own calculations.

Table 25: ET and additional costs of the chemical production, 2010 and 2020, Mio. €

But also initially less carbon intensive productions get an additional cost pressure, as seen for the Non-iron metal production in table 23 and for the sector pulp and paper in table 24.

Furthermore, additional burdens can be expected for sectors which are not included in the ET scheme. For instance the chemical production, see table 25, which is, on account of its relatively high power intensity, specially affected by the price increase for electricity. The ET scheme is similar to the green tax reform: the scheme will induce a stronger spreading of sectoral production costs and thereby accelerate the sectoral change beyond the CO_2 reductions. Therefore the energy intensive production will be replaced by less energy intensive production. Additionally, domestic production will be replaced by imports from third countries, which can be produced in less efficient installations (leakage).

6.5 Macro economic effects

The effects will not be limited to the sectors included in the ET scheme but will also affect other sectors by cost transmissions. Thereby price increases are caused in the non-energetic area of intermediates, consumption, investment, and exports.

In principle, the cost transfer process depends on the demand and supply conditions and will therefore differ between the sectors regarding extent and duration. Relevant for the concrete effect is, above all, the competition situation on each market. A strong market position of import products limits the transfer of rising costs of the domestic production upon the prices. Above all, this applies to all energy intensive mass products which are produced by standardised techniques, for instance cement, glass, steel, and paper. This applies even if the EU proposal causes burdens for all member states, as the system harms the competitiveness towards suppliers from Non-EU countries.

The cost impulse will not only harm the competitiveness of energy intensive sectors but will also transfer burdens to the energy extensive sectors of manufacturing. Accordingly, this mechanism will cause job losses of about 100,000 until 2012, which will not occur in the more flexible scheme integrating JI and CDM, see table 26. The German employment situation will be harmed above-average. Altogether, the statement of the European Commission that the EU-wide ET scheme on the level of companies respectively installations of the energy sector and the energy intensive production will not harm employment is wrong. The statement bases on the hazardous misunderstanding of model simulations and real economic adjustment processes: in an Applied General Equilibrium model of a type which was taken by the EU, structural adjustment processes, regional rejections, and frictions on the labour market, do not exist per assumption. In reality, these adaptation restraints have to be considered above all, if a new instrument is to be implemented.

"CO₂ Emissions Trading put to test
– Design problems of the EU draft directive concerning an Emissions Trading scheme"

	ET incl. JI and CDM		EU proposal	
	GDP in %	Employees in 1000	GDP in %	Employees in 1000
		2010		
Belgium	-0.02	-0.6	-0.10	-4.0
Denmark	-0.00	-0.1	-0.02	-0.5
Germany	-0.01	-4.0	-0.07	-27.6
Finland	-0.02	-0.5	-0.18	-4.0
France	-0.01	-1.4	-0.04	-8.6
Greece	-0.01	-0.6	-0.10	-3.8
Ireland	-0.01	-0.1	-0.04	-0.7
Italy	-0.01	-2.4	-0.06	-14.3
Netherlands	-0.01	-1.2	-0.09	-7.3
Austria	-0.01	-3.0	-0.05	-18.3
Portugal	-0.01	-0.7	-0.08	-3.9
Sweden	-0.01	-0.2	-0.03	-1.4
Spain	-0.01	-1.7	-0.07	-11.6
UK	-0.01	-2.0	-0.04	-11.4
EU 15 total	-0.01	-15.6	-0.06	-100.7
		2020		
Belgium	-0.02	-0.6	-0.11	-4.3
Denmark	-0.00	-0.1	-0.02	-0.6
Germany	-0.01	-3.6	-0.09	-34.3
Finland	-0.02	-0.4	-0.26	-5.8
France	-0.01	-1.4	-0.04	-9.0
Greece	-0.01	-0.5	-0.12	-4.6
Ireland	-0.01	-0.1	-0.04	-0.7
Italy	-0.01	-2.3	-0.07	-15.0
Netherlands	-0.01	-1.1	-0.09	-7.4
Austria	-0.01	-3.0	-0.05	-18.3
Portugal	-0.01	-0.7	-0.08	-3.9
Sweden	-0.01	-0.2	-0.03	-1.4
Spain	-0.01	-1.6	-0.08	-12.0
UK	-0.01	-1.8	-0.04	-11.4
EU 15 total	-0.01	-14.6	-0.07	-112.7

Own calculations.

Table 26: Macro economic effects of ET, 2010 to 2020

7 Conclusion

7.1 The EU Draft Directive from 23. October 2001

As criteria for an evaluation of climate policy instruments like Emissions Trading one has to count conceptual feasibility and the impact in terms of a Sustainable Development. In this connection, it is necessary to fulfill both criteria. It cannot be the aim to realise a political concept which contradicts several laws, other political goals, or economic mechanisms. At the same time, it cannot be reasonable to realise a feasible concept if the expected effects regarding ecological, economic, and social aspects turn out negative.

It has been shown, that conceptual problems for the implementation of an ET scheme in the EU were caused by the fact that only a part of the world committed itself to reduce their carbon emissions. Therefore, the theoretical basic concept of Emissions Trading cannot be realised unmodified without leading to ecological and economic harmful leakage effects in countries without reduction commitments. As this basic problem was seen on the UN level, the flexibility mechanisms were central aspects of the Kyoto Protocol and the following negotiations. Therefore, an EU-wide implementation of the covenants made on the UN level should also include the participation in Annex B-wide ET and the assessment of project based emission reductions.

The EU Draft Directive from 23. October 2001 ignores the basic problem in spite of the ecological, economic and social relevance by defining a classic Cap-and-trade system on the level of installations with absolute caps. The analysis of the conceptual level has shown that the EU Draft Directive has to be rejected for its internal caveats. Additionally, it can be stated that the rejection is caused by the implementation of absolute caps on the level of installations respectively companies. The conceptual problems range from questions about the initial allocation to contradictions regarding competition regulations.

The practicability of auctions is discussed controversially on the juridical level. Politically, there seem to exist preferences for partial auctioning, at least in the EU parliament. For the impact of (partial) auctioning in the international competition the (partial) auctioning would lead to a (partial) expropriation on account of the resulting burden. Additionally, with every initial allocation against payment, the question about unanimity on the EU level comes up.

The second politically discussed possibility is Grandfathering. But Grandfathering itself holds several serious problems if it is connected with absolute caps. The fundamental request of competition neutrality for the initial allocation and the following production effects (i.e. capacity fluctuations, newcomers) cannot be realised with the acceptance of early actions on the basis of absolute caps and the great differences of member states commitments (BSA). The contradictions lead to several subsequent economic and competitive problems. These problems cannot be solved on the basis of absolute caps. The Draft Directive reflects most of the contradictions without showing a method of solution. Instead, the problematic and contradictive aspects are covered by scopes of interpretation which are competitively very questionable.

Therefore, not only the EU Draft Directive, but all ET schemes have to be rejected according to the conceptual level which integrates installations respectively companies on the basis of absolute caps. This applies also in the case of very low price effects, which can be reached by instrumental modifications. The contradictions remain, their explicit appearance will only be set back temporally to a later period when the price for CO_2 will rise. In the past, in several areas mechanisms were implemented, like in the European agricultural policy, educational policy and labour market policy, which caused negative long-term effects due to their wrong conception. It is obvious, by looking at current negotiations, how difficult subsequent modification attempts are. Thus, a short-term designed European climate policy has to be refused.

Also on the level of forecasting impacts, the EU Draft Directive has to be evaluated negatively. The risk of high leakages into non-restricted countries, which was realised on the UN level, is confirmed by model simulations (chapters 5 and 6). The strict defined EU Draft Directive leads to the highest leakage (20.1%) at CO_2 price of 12.70 $/t. This price is less than the EU forecast, but it implements much higher economic and social costs. The higher the leakage results the more employment will be lost in regions with emission restrictions. Relocations will occur first in sectors with the highest reduction costs and where companies have to compete internationally and cannot transfer the cost rise onto the market. The EU calculations in their model, abstract from possibilities of re-locations and thus deny the existence of leakage; Therefore, those companies also buy permits on the internal EU permit market who, in reality, will not be able to. Thus, the lower CO_2 price is connected with a higher loss in employment.

An implementation of the flexible mechanisms from Kyoto will lead to a significant decrease in costs and a broad enhancement of ecological impacts. The price for CO_2 drops to 0.33 \$/t, which has to be classified as a temporary effect of the existing Hot Air amount. Regarding the forecasted growth in Russia, the Hot air will be depleted until 2018 and the CO_2 price will range about 5 \$/t CO_2. The leakage in this scenario will be about 4.7% and the global emissions will be reduced by 1.3%. In comparison, the EU Draft Directive will only cause a reduction of about 0.8%. The reductionof global emissions has to be seen as the more important political goal to fight global warming than the cleansing of national GHG accountings. Otherwise the EU will export their CO_2 emissions and import the produced goods. These results will not change in principle, if Russia will supply their permits strategically.

It can be concluded from the model scenarios that a flexible concept will have positive ecological effects. Yet, this should not conceal conceptual defects. The aim must be the constructive attempt to work out an alternative, that promises an ecologically, economic, and socially acceptable solution. Such a concept have to be deduced from the UN level, must avoid conceptual contradictions, and fulfill the shown criteria of competition neutrality. At the same time, the conception has to be elaborated with the help of the model scenarios, that ecological effects are maximised and simultaneously economic, structural, and social harms are avoided. Thus, the EU Draft Directive has to be rejected on the base of Sustainable Development considerations for its ecological, economic, and social impact.

7.2 The Danish compromise proposal from 28. August 2002

On 28.08.2002 the European Commission presented under the Danish presidency, presented a modified proposal with the reference number ENV/02/8. This proposal differ from the EU Draft Directive from 23.10.2001 in some important points. The negative impacts caused by the implementation of an isolated ET scheme in the EU were realised. Therefore the flexible mechanisms were integrated differently in the Danish proposal.

In comparison to the Draft Directive the following points were modified:

- Widening of the included activities by Article 3c,

- Widening of the possibility to include more installations (Opting in) and other GHG (Article 23 bis),

- Mandate to conclude contracts with other Annex B-countries to allow a mutual acceptance of both ET schemes (Article 24),

- Possibility for an opting out for special installations if they fulfill strict conditions until 31.12.2007,

- Implementation of JI and CDM from 2005 on,

- Emissions Trading in terms of the Kyoto Protocol will be taken into consideration but is not defined as a legal claim,

- Modifications in details about sanctions, acknowledgement of permits from other ET schemes of Annex B-countries, and banking,

- Modification of the allocation criteria, specially, omission of the formula that no installation is allowed to get more permits than needed, moreover, a restriction of the transfer of reductions from measures in the period from 2005 to 2007 into the period of 2008 to 2012.

If the voluntary agreement of the German industry can be made compatible at least until 2007 is questionable, according to the request for comparable sanctions and the reserved authorisation by the European Commission. From 2008 on the lack of instrumental compatibility will occur anyhow. Temporal synchronisation with the Kyoto mechanisms is not guaranteed.

Therefore the Danish compromise improves the Draft Directive in reference to the needed flexibilisation by implementing the Kyoto mechanisms. But the fundamental conceptual problems remain unsolved.

7.3 Net result

International climate policy constituted with the Kyoto Protocol and its three instruments, the best possible political framwork. These instruments have to be used in European climate policy. As these national instruments were able to reduce emissions signficantly they should not be harmed. Every national and European policy has to start from the international climate policy framework and consider the criteria from chapter 3:

- Complete integration of all instruments of the Kyoto mechanisms

- Temporal synchronisation with the Kyoto period 2008 – 2012
- In principal, no auctioning of permits (not even partially)
- Consideration of all early actions since 1990
- No competitive distortions within the EU and a minimisation of effects towards third countries, that means that, for instance, a coal fired power plant with a comparable efficiency has to be treated equally in Spain, Ireland, and Germany.

De facto, the EU Draft Directive from 23.10.2001 does not fulfill any one of these criteria. Thus, the Draft Directive does neither fulfill the first main requirement of conceptual feasibility nor the demand for an outlook for a best efficient fulfilment of the aims of Sustainable Development. The Draft Directive is therefore no reasonable framwork for the climate policy in Europe and Germany. The Danish compromise from 28.08.2002 recognises some of the disputed aspects and is in so far an improvement. But the remaining basis of absolute caps for installations shows no conceptual progress in terms of, for instance, competitive neutrality.

The European Commission is still invited to consider the Draft Directive thoroughly. The Kyoto mechanisms and the existing national instruments represent the base for further political suggestions. If at all, Emissions Trading seems to be a reasonable instrument for German industry just in the long term.

Therefore, the EU Draft Directive from 23. 10. 2001, as well as the current Danish compromise cannot be accepted.

Literature

Arndt, H.-W., B. Heins, B. Hillebrand, E.-C. Meyer, W. Pfaffenberger and W. Ströbele (1998), Ökosteuern auf dem Prüfstand der Nachhaltigkeit, *Angewandte Umweltforschung*, No 13, Berlin 1998.

Birnbaum, K.U., R.Pauls, H.-J.Wagner, M.Walbeck (1991): Berechnung sektoraler Kohlendioxidemissionen für die Bundesrepublik Deutschland, in: Jülich (2530), Reihe Angewandte Systemanalyse, No 62, 1991.

Blok, K. et al (2001): Economic Evaluation of Sectoral Emission Reduction Objectives for Climate Change – Comparison of ,Top-down' and ,Bottom-up' Analysis of Emission Reduction Opportunities for CO_2 in the European Union, Memorandum 2001.

Blok, K., D. de Jager, C. Hendriks (2001): Economic Evaluation of Sectoral Emission Reduction Objectives for Climate Change – Summary Report for Policy Makers, updated, March 2001.

Böhringer, C., T. F. Rutherford (2000): Decomposing the Cost of Kyoto – A global CGE Analysis of Multilateral Policy Impacts, ZEW Discussion Paper No. 00-11.

Böhringer, C. (2002): Climate Politics from Kyoto to Bonn: From Little to Nothing?, Energy Journal, Vol. 23, No. 2, S. 51-71.

Brockmann, K.L., M. Stronzik, K.Bergmann (1999): Emissionsrechtehandel – eine neue Perspektive für die deutsche Klimapolitik nach Kioto, Heidelberg.

Bundesministerium für Wirtschaft und Technologie (2001): Nachhaltige Energiepolitik für eine zukunftsfähige Energieversorgung, Energiebericht, Berlin, October 2001.

Bundesministerium für Wirtschaft und Technologie (2000): Energie Daten 2000, Bonn, 2000.

Burniaux, J.-M., Oliviera Martins, J., (2000): Carbon Emission Leakages: A General equilibrium View, OECD Economic Department Working Paper No. 242, Paris.

Buttermann, H.G. (1999): Potenziale und Kosten einer globalen CO_2-Minderungsstrategie in der Stahlindustrie, RWI-Mitteilungen, 1999, 50(3), pp 145.

Buttermann, H.G., B. Hillebrand (2002a): Die Klimaschutzerklärung der deutschen Industrie vom März 1996 – eine abschließende Bilanz, Monitoring-Bericht 2000; Untersuchungen des Rheinisch-Westfälischen Instituts für Wirtschaftsforschung, No 40, Essen, 2002.

Buttermann, H.G., B. Hillebrand (2002b): Sektorale und regionale Wirkungen von Energiesteuern, (Untersuchungen des Rheinisch-Westfälischen Instituts für Wirtschaftsforschung, No 31), Essen, in preperation.

Cansier, D. (1993): Umweltökonomie, Stuttgart.

Capros, Pantelis/ Leonidas Mantzos (2000): The Economic Effect of EU-Wide Industry-Level Emission Trading to reduce Greenhouse Gases – Results from PRIMES Energy System Model, Athens 2000.

Capros et al (w/o year): The *PRIMES* Energy System Model – Reference Manual, Athens without year.

Commission of the European Communities (2001): Proposal for a Directive of the European Parliament and of the Council establishing a framework for greenhoues gas emissions trading within the European Community, Brussels, 31.05.2001.

Dieckheuer, G. (2001): Internationale Wirtschaftsbeziehungen, 5. Edition, Munich.

European Union (2000): Green Paper on greenhouse gas emissions trading within the European Union, presented by the Commission, COM(2000),87 of March 2000.

European Parliament (2002): Draft report on the provisional for a European Parliament and Council directive on establishing a scheme for greenhouse gas emission allowance trading within the Community and amending Council Directive 96/61/EC, COM(2001) 581 – C5-0578/2001 – 2001/0245(COD).

Feess, E. (1995): Umweltökonomie und Umweltpolitik, Munich.

Hensing, I., W. Pfaffenberger, W. Ströbele (1998): Energiewirtschaft, Munich.

Jäger, G., K.A. Theis (2001): Increase of Power Plant Efficiency, VBG PowerTech, 2001, issue 11, pp 21.

Jochem, A. (1999): Rahmenbedingungen für ein internationales System handelbarer Emissionsrechte im Kyoto-Protokoll, Zeitschrift für Umweltpolitik und Umweltrecht, 22 (3), pp 349-368.

Light, M. (1999): Coal Subsidies and Global Carbon Emissions, Energy Journal, 20, No. 4, pp 117-148.

McKibbin, W.J., P.J. Wilcoxen (2002): The Role of Economics in Climate Change Policy, Journal of Economic Perspectives, 16, No 2, 2002, pp 107-129.

Klemm, A. (2002): Klimaschutz nach Marrakesch, Heymanns, Cologne et al.

Metzger, B.R., A.Pelchen (2001): CO_2-Emissionsrechte: Was bringt ein Handelssystem für ein Unternehmen ?, Elektrizitätswirtschaft (ew), 100, No 10, pp 34-39.

Neumann-Mahlkau, P. (2002): Treibhaus oder Kühlhaus ? – das Klima der Erde, Energiewirtschaftliche Tagesfragen, 2002, No 1 / 2, pp 2.

Painuly, J.P. (2001): The Kyoto Protocol, Emissions Trading and the CDM: An Analysis from Developing Countries Perspective, The Energy Journal, 22, No 3, pp 147-169.

Paltsev, S.V. (2000): The Kyoto Agreement: Regional and Sectoral Contributions to Carbon Leakage, Working Paper, Center for Economic Analysis, Department of economics, University of Colorado at Boulder.

Peters, M., P. Schmölde (2002): Einblasen von Ersatzreduktionsmitteln in den Hochofen - Auswirkungen auf Metallurgie und Kosten, „Stahl und Eisen", 122 (2002), No 4, pp 43.

PROGNOS AG (2001): Energiepolitische und gesamtwirtschaftliche Bewertung eines 40 %-Reduktionsszenarios, herausgegeben vom Bundesministerium für Wirtschaft und Technologie, Documentation No 492, 2001.

Roeder, G. (2002): Emissionshandel aus Sicht eines energieintensiven Unternehmens, VIK-Mitteilungen, No 2.

Rühland, P.-C.(1993): Moderner 500-Megawatt-Steinkohleblock im Kraftwerk Staudinger, Energiewirtschaftliche Tagesfragen, 43 (1993), pp 624

RWE-Rheinbraun (Eds.): Braunkohle in Europa, 2001, Cologne 2001, pp 12.

Schafhausen, F.-J. (2002): Der Kampf um die Ratifizierung des Kyoto-Protokolls in Marrakesch, Energiewirtschaftliche Tagesfragen, No 1 / 2.

Schafhausen, F.J. (2002): Der Emissionshandel als klimaschutzpolitisches Instrument, Energiewirtschaftliche Tagesfragen, No 8.

Schärer, B. (2001): Erheblicher Erkenntniszuwachs beim Klimaschutz – Bericht der Arbeitsgruppe Minderungsmaßnahmen des IPCC, Elektrizitätswirtschaft (ew), 100, No 10, pp 18-24.

Smajgl, A. (2001): Modellierung von Klimaschutzpolitik – Ein Leitfaden; Ressourcen Journal, No 3, 2001.

Smajgl, A. (2002): Modellierung von Klimaschutzpolitik: Ein Allgemeines Gleichgewichtsmodell zur ökonomischen Analyse der Wirkung von CO_2-Restriktionen auf den Einsatz fossiler Energieträger, LIT-Verlag Münster 2002.

Smajgl, A., W. Ströbele (2001a): Zertifikatshandel für Treibhausgase? Anmerkungen zum EU-Vorschlag vom März 2000, Volkswirtschaftliche Diskussionsbeiträge No 321, 2001.

Smajgl, A., W. Ströbele (2001b): Modeling carbon abatement policy and the optimal supply of exhaustible resources, forthcoming, 2001.

Ströbele, W., H. Wacker (2000): Außenwirtschaft, 2. Edition, Munich.

Sun, J.W. (1999): Decomposition of Aggregate CO_2 Emissions in the OECD: 1960 – 1995, The Energy Journal, 20, No. 3, pp 147-155.

Sutherland, R.J. (2000): No Cost Efforts to Reduce Carbon Emissions in the U.S.: An Economic Perspective, The Energy Journal, 21, No. 3, pp 89-112.

VEAG (2000), Geschäftsbericht 1999, Berlin 2000.

VIK (2001): Statistik der Energiewirtschaft 2000/2001, Essen 2001.

w/o author (2000): 1000-Megawatt Spitzenkraftwerk am Netz, Verbund 2000, No 2, Essen.

Ziesing, H.J. (2000): CO_2-Emissionen im Jahre 1999: Rückgang nicht überschätzen, DIW-Wochenbericht, No 6/2000.

Annex 1: code TM – The Applied General Equilibrium model

Model aggregation

- Division of the world economy in 13 regions

OCE	Australia and New Zealand
JPN	Japan
NAA	Non-Annex B-Asia
CAN	Canada
USA	United States of America
RAM	Rest of America
UKM	United Kingdom
DEU	Germany
REU	Rest of Western Europe
CEA	Central Europe and Turkey
FSU	Former Soviet Union
MEN	Middle East and North Africa
RWO	Rest of the World

- Division of each economy in 13 sectors

COL	Hard coal
BCL	Brown coal and lignite
OIL	Crude oil
GAS	Natural gas
P_C	Oil products
ELY	Electricity
ISM	Iron, steel, and non-iron metals
SCG	Stones, ceramic and glass
PPP	paper
CHE	Chemistry and pharmaceutical industry
T_T	Transport
CGD	Savings
OTS	Other sectors

The model structure

The AGE model *CODE* **TM**[70] (*Climate policy scenarios in a dynamic general equilibrium trade model*) analyses the impact of carbon restrictions set by the UNFCCC agreements. The world economy is aggregated in 8 sectors and 16 regions. The model is defined as a mixed complementarity problem.[71] It is programmed in GAMS MPSGE[72] and uses GTAP 4E data.[73]

The production Y_r of region r is defined as

$$Y_r = \vartheta_r \cdot \left(\theta_K \cdot K_r^{\rho_Y} + \theta_L \cdot L_r^{\rho_Y} + \theta_R \cdot E_r^{\rho_Y} + \theta_N \cdot N_r^{\rho_Y}\right)^{1/\rho_Y} \qquad (M.1).$$

K_r, L_r and E_r represent capital, labour, and the energy aggregate. N_r denotes the intermediate input, ϑ_r represents the overall productivity of a region, and $\sigma_Y = 1/(1-\rho_Y)$ stands for the elasticity of substitution. The domestic production can either be exported or consumed. The latter part is combined with imports so that the sectoral Armington good results.

$$A_r = \left(\left(\theta_Y \cdot Y_r - \theta_{XP} \cdot XP_r\right)^{\rho_A} + \theta_{MP} \cdot MP_r^{\rho_A}\right)^{1/\rho_A} \qquad (M.2)$$

The sum of world-wide exports in a sector must be equal to the sum of its imports.

$$\sum_r XP_r = \sum_r MP_r \qquad (M.3)$$

The Armington good allows private (C_r) and public consumption (G_r).

According to the Heckscher Ohlin assumption crude oil is modelled as a world-wide homogeneous good, as is coal with regard to its energy content. Natural gas, which is delivered by pipeline, is treated as a continental homogeneous good. As a consequence, within an existing pipeline network, the con-

---

[70] A detailed description of *CODE* **TM** can be found in Smajgl (2002).
[71] Cf. Rutherford (1995a) and Flakowski (2002).
[72] Cf. Brooke/Kendrick/Meeraus (1992) and Rutherford (1995b).
[73] Cf. McDougall/ Elbehri/ Truong (1998); Malcolm/Truong (1999); Complainville/van der Mensbrugghe (1998);

sumer has no preference for any supplier. For LNG global homogeneity is assumed.

$$A_r = \left(\theta_C \cdot C_r^{\rho_C} + \theta_G \cdot G_r^{\rho_C}\right)^{1/\rho_C} \quad (M.4)$$

The energy aggregate in (M.1) is nested on four levels. Extracted crude oil and natural gas appear on the lowest level.

$$OG_r = \vartheta_r^{OG} \cdot \left(\theta_{OIL} \cdot OIL_r^{\rho_{OG}} + \theta_{GAS} \cdot GAS_r^{\rho_{OG}}\right)^{1/\rho_{OG}} \quad (M.5)$$

This aggregate OG_r can be substituted for coal on the next level and produces the aggregate OGC_r. OGC_r enters the next level with electricity ELY_r and creates EN_r. On the last level EN_r can be substituted for refined oil products OPR_r, and they create the energy aggregate E_r which enters (M.1).

$$OGC_r = \vartheta_r^{OGC} \cdot \left(\theta_{OG} \cdot OG_r^{\rho_{OGC}} + \theta_{COAL} \cdot COAL_r^{\rho_{OGC}}\right)^{1/\rho_{OGC}} \quad (M.6)$$

$$EN_r = \vartheta_r^{EN} \cdot \left(\theta_{OGC} \cdot OGC_r^{\rho_{EN}} + \theta_{ELY} \cdot ELY_r^{\rho_{EN}}\right)^{1/\rho_{EN}} \quad (M.7)$$

$$E_r = \vartheta_r^{E} \cdot \left(\theta_{EN} \cdot EN_r^{\rho_E} + \theta_{OPR} \cdot OPR_r^{\rho_E}\right)^{1/\rho_E} \quad (M.8)$$

The dynamic bases on a steady state path with the regional growth rates g_r. The basic dynamic equilibrium condition for the Ramsey path is defined by $r = g + \delta$.[74] The underlying maximisation of the utility is represented by

$$\max \sum_{t=1}^{T} U(C_t) \cdot (1+\delta)^{-t} \quad \text{with} \quad U(C) = \ln C, \quad (M.9)$$

s.t. $k_j^t = K_j^t - K_j^{t-1}$.

---

[74] This does only apply if the elasticity of marginal utility (the reciprocal of the intertemporal elasticity of substitution) is $\eta = 1$, which is met by $U(C) = \ln C$, because $\hat{C} = \dfrac{r-\delta}{\eta} = g$ with exogenous interest rate r and preference rate δ.

Modeling fossil fuels

Substitution is not only caused by carbon restriction but also by the exhaustibility of fossil fuels. This fact is frequently neglected in AGE models. There exists a vast body of literature about the scarcity of fossil fuels and the existence of a Backstop technology in dynamic optimisation models, see Hoel/Kverndokk (1996) and Tahvonen (1997). The GREEN Model provides an almost endogenous technique to consider the scarcity of fossil fuels in an AGE approach, see Burniaux et al. (1992) and OECD (1994). CODE TM offers a new approach of modeling the interplay of the fossil fuel scarcity and the scarcity of the atmosphere. This mechanism is derived from an optimisation approach which is extensively presented in Blank/Ströbele (1995) and Smajgl/Ströbele (2002). The maximisation problem is defined as

$$\max \int_0^\infty \left(e^{-\delta \cdot t} \cdot U(R + BST) - k \cdot BST \right) \cdot dt \qquad (R.4)$$

s.t. $\dot{S} = -R$ with $S \geq 0$ and given S_0 (R.5)

$\dot{M} = -\gamma \cdot M + R$ with $M \leq \overline{M}$ and given M_0 (R.6)

The Hamiltonian of this modified problem is:

$$\tilde{H} = U(R + BST) - k \cdot BST + \mu \cdot (-R) - \lambda \cdot (-\gamma \cdot M + R). \qquad (R.7)$$

Solving the problem yields the following results: Marginal utility equals the sum of the shadow prices:

$$U'(R) = \mu + \lambda \qquad (R.8)$$

The paths of shadow prices are

$$\frac{\partial \tilde{H}}{\partial S} = 0 = -\dot{\mu} + \delta \cdot \mu \quad \Rightarrow \quad \mu = \mu_0 \cdot e^{\delta \cdot t} \qquad (R.9)$$

$$\frac{\partial \tilde{H}}{\partial M} = \lambda \cdot \gamma = -\dot{\lambda} - \delta \cdot \lambda \quad \Rightarrow \quad \lambda = \lambda_0 \cdot e^{(\delta + \gamma)t} \qquad (R.10)$$

Because λ is associated with a *bad* it reduces utility. Simultaneously, the resource supplier's Hotelling rent decreases. The connection between optimal supply behaviour on fossil fuel markets and GHG mitigation policy is obvious. Therefore it is necessary to implement a scarcity of fossil fuels. Resource theory presents the optimal solution in a first best situation with perfect foresight and perfect capital markets.[75] Theoretically, resource stocks are depleted according to their extraction costs, starting with the lowest costs.

However, in the real world some assumptions of the presented simple Hotelling type model cannot prevail. Since the big oil companies lost their property rights in the mid seventies they chose to explore new oilfields outside the OPEC-area. For instance, significant contribution to the world oil market was created by drilling oil in Alaska and the North Sea. Of course, production costs there were much higher than in the Middle East.

Since then, the game has evolved into an interplay between long-term behaviour of OPEC and big oil companies, and was intensified by the disappearing of GULF or the fusion of EXXON and MOBIL. However, both players have to cope with a group of fringe suppliers. As coal and gas substitute for oil in large market segments, crude oil prices will not follow the Hotelling path predicted by theory.[76] These game theoretical aspects are modelled in, for instance, Berg/Kverndokk/Rosendahl (1997 a/b).

Nevertheless, scarcity of fossil fuels has an important impact on the fossil fuel's prices and therewith on the GHG price. To model climate policy scenarios appropriately, it is necessary to implement the carbon restriction as well as the Hotelling rule and, in a further step, the complementary (game theoretical) factors mentioned above. In this analysis the interplay of the scarcity of the atmosphere and the scarcity of crude oil is implemented.[77]

This analysis is limited to a time period ending in 2012. Therefore an explicit modeling of a Backstop technology is omitted and only crude oil is implemented as a scarce resource. Because of the short time horizon the use of crude oil is considered as a flow without modeling the resource stocks:

---

[75] In this paper the theoretical background can only be summarised. For a deeper insight see Hanley/Shogren/White (1997).
[76] Otherwise the strong reluctance of oil exporting countries like Saudi-Arabia to a strong climate change policy could not be understood.
[77] See Smajgl (2002) or Smajgl/Ströbele (2002) for a detailed methodical description.

$$\pi_{R,t} = HR_t + RR_{R,t} = \left(p_t^{OIL} - c_{R,t}^{OIL}\right) \tag{R.1}$$

$$\text{with} \quad HR_t = HR_0 \cdot e^{-r \cdot t} \tag{R.2}$$

$$OIL_0 = OR_0 + RR_0 + HR_0 \tag{R.3}$$

OR indicates the production costs themselves, and the intramarginal rent is divided into a part that could rise with the interest rate (HR: Hotelling rent) and a part which is not sensitive to the interest rate, the latter one is called Ricardian rent (RR). This assumption is necessary as different regions extract crude oil simultaneously although they have different extraction costs.

Simultaneous production from oil wells with different production costs $c_{R,t}^{OIL}$ in combination with a uniform market price for homogeneous goods, crude oil implies that the intramarginal rents $\pi_{R,t}$ of different suppliers will vary, see (R.1).

If these different intramarginal rents all increased with the interest rate (R.2) the region with the lowest production costs would have the strongest price increase over time and consequently its price would overtake the price of other suppliers. Following the assumption that crude oil is homogenous,[78] this region could not sell a single barrel according to the Hotelling rule. The extraction costs of North Sea producers vary between $8-9 /bbl, while those of Saudi Arabia are between $0.5-2 /bbl. Following the optimal Hotelling path, the Saudi oil is obstructed by North Sea oil production.

As soon as the deposits with the highest production costs are exhausted, the other suppliers will shift portion of the intramarginal rent from RR to HR, increasing the price path's slope. In a full dynamic approach the path of HR_t must follow the Hotelling rule.

---

[78] Of course, crude oil is not homogenous in a technical sense; positive or negative surcharges are, however, a suitable economic practice due to quality differences, different refinery costs, transportation costs etc..

Annex 2: The RWI model system

Model structure

The models developed at the Rhine-Westphalian Research Institute of Economics focus on the implementation of interactions between energy economic, sectoral and macro economic developments. It arose by the experiences of the two energy price increases in the mid and the end of the seventies, which caused significant sectoral and price impulses including macro economic distortions. In the last years the model was enlarged by ecological questions. The model is used to forecast developments on the energy level and macro economics as well as to simulate the impact of decisions of economic policy, energy policy and environmental policy. For this purpose three parts can be identified:

> The energy model includes the different steps of energy supply from primary energy to the final energy demand following the patterns of the energy accounting scheme;

> The structural model is differentiated into 60 sectors following the patterns of Input Output schemes and displays the real good demand and the price formation in a sectoral structure;

> The environmental model is used to calculate the external effects from consumption and transformation of energy, including surface sealing, and the costs of significant reduction and avoidance techniques.

Energy model

The aim of the energy model is the realistic explanation of energy demand and energy supply, and the calculation of energy costs as a part of sectoral cost structures and as a part of total expenses of private households. This aim is reflected by the formal model structure, in the choice of the modelling approach and the level of aggregation. As the sectoral organisation corresponds to the system of Input Output calculations and the differentiation in fourteen energies corresponds to the system of an energy balance, the energy input in energy intensive sectors is significantly stronger disaggregatedly modelled. Even if the sectors have specific consumption conditions the energy consumption is formally designed similarly. It results in the multiplication of components for outfit and specific consumption and utility. The first component stands for the capital stock which exists at a special point of time. The second one explains the technical design of capital goods, and the third one its economic utility.

The supply side of the energy model is used to cover the primary energy needed for consumption of final energy consumption and non-energetic use. In this context, the model traditionally distinguishes between the transformation of hard and brown coal in coke and briquette production, the distillation of crude oil in refineries, the power generation in public and industry power plants, and the production of district heating in heating power plants and long-distance heating plants. At the same time, the energy prices are determined for each energy good and consumer group by the costs which arise in the listed transformation areas.

Structural model

The structure model can be characterised as a completely integrated system to explain sectoral and macro economic developments. Following the patterns of interlocking schemes of an Input Output table the following modules can be identified:

- an Output system for the commodity markets in which final demand and intermediates are implemented;

- a price model in which the sectoral unit costs, consisting of different cost components (labour, capital, intermediates) and taxes respectively subsidies, are explained and prices of each commodity respectively commodity group are calculated;

- a calculation of the total capital stock and its potential in which sectoral gross outfit and gross building assets are explained,

- a labour market model in which the labour supply is calculated on the macro economic level and on the sectoral level the amount of labour, the average labour times, the number of employees, the labour productivity, wages and gross income from employed labour is explained,

- a redistribution model which allows the implementation of the tax system, social insurances, and transfers, and therefore the calculation of public incomes and expenses and the disposable income of private households.

Environmental model

The environmental model aspires to balance all ecological impacts of energy consumption and energy transformation. Thus, the calculation has to implement fuel combustion caused by traditional gases like CO and NO_x, Greenhouse Gases like CO_2 and CH_4, the unused heat loss caused by the transformation of primary energies in secondary energy and by the transformationof final energy in utility energy, and specially the radioactive and non-radioactive waste in power generation. Additionally, the area needed for settlement and traffic infrastructure, and indicators for energy and raw material efficiency of the German industry are taken into account.

By balancing the physical flows of environmental influences the environmental model includes several costs which describe the additional investments for techniques for emission reductions or the avoidance of whole bundles of emissions. This specially includes renewable energies in power generation and in the area of private households and small demand, reduction costs of energy input for heating, and, lately, also techniques to reduce environmental influences in the traffic sector (methanol drive, fuel cell on the basis of hydrogen or natural gas, Eco diesel).

Simulation possibilities

For the detailed implementation of the energy sector and specially of the energy prices all possible variations of additional taxes on energy consumption, energy supply or the sealing of surface can be forecasted. For instance, the green tax reform from 1. April 1999. Restrictions are caused by the sectoral aggregation of 60 sectors. For instance, the specific burdens and disburdens of the green tax can only be calculated in recourse on a significant amount of detailed information on the basis of four digit commodity group numbers. The sub-structure of capital stocks in vintages allows simulations which are usually not given in economic models. For instance the impact of consumption taxes for new registered cars on the traffic and its emissions, the effect of directives for heating standards for buildings or heating installations on the demand for room heat, or the subsidies for renewable energies for power generation and the utilisation. The connection of changes in capital stocks and investment of the sectoral model and the energy costs in sectoral cost structures guarantees for all simulations the consideration not only of energy economic and ecological effects but also of the macro economic impact.

Glossary

Bottom up model — This type of model tries to simplify reality by implementing the most important economic sectors in a very detailed way. For instance to implement the energy sector on the level of single installations.

Annex B countries — This group of countries was defined in Annex B of the Kyoto Protocol as those countries with reduction commitments.

Banking — The Kyoto Protocol defined this instrument to allow flexibility emission reductions in the temporal dimension and, therefore, the possibility to transfer a surplus of generated reductions into later periods.

Burden Sharing Agreement — The EU member states agreed in this contract on a disproportional distribution of the Kyoto reduction commitment of 8% between each other.

Business as usual — To allow a judgement of different policy scenarios a reference case is needed, the so called *Business as usual* (BAU). It adjusts the current situation in the future and by that delivers a basis for the relative evaluation.

Cap and trade system — Such an approach limits the amount of allowed emissions (*cap*) and makes the permits tradable (*trade*).

Clean Development Mechanism — The Kyoto Protocol defines this instrument as allowing a regional flexibility for emission reductions through the possibility to realise reduction projects in Non-Annex B countries and to lower the own commitment by these reductions.

Compliance — Constitutes the system to guarantee the fulfillment of commitments by monitoring, reporting, verification, and sanctioning.

Emissions Trading — The Kyoto Protocol defines this instrument as allowing a regional flexibility for emission reductions through the possibility to buy permits from other

	Annex B-countries and to lower the own commitment by these permits.
Host countries	Countries where emission reduction projects (*Joint implementation* und *Clean Development Mechanism*) are realised.
Hot air	The amount of supplied permits which represent already achieved emission reduction from decreases in production level, i.e. in the Former Soviet Union.
Hot spot problem	Environmental problems which are regionally localised without a global or international effect.
Joint Implementation	The Kyoto Protocol defines this instrument to allow a regional flexibility for emission reductions by the possibility to realise reduction projects in Annex B countries and to lower the own commitment by these reductions.
leakage	Index for the ecological efficiency of an instrument to fight global warming. It is defined by the relation between increases in emissions of countries without a reduction commitment and emission reductions in countries with commitments. A leakage of 20% means that 20% of the success of the restrictive countries are offset by production shiftings into regions without reduction commitments.
Merit order	Describes the sequence of use of powerplants in the base load, middle load and peak load, and therefore of the input of fossil fuels, by the amount of marginal costs.
Monitoring	Describes the survey of emissions by a neutral institution.
Top down model	This type of model simplifies reality by defining the production of the national economy with its international trade linkage and by implementing the important sector in a less aggregated way. (In comparison: Bottom up model)
Wall fall profits	Collective term for the emission reductions of about 100 Mio. t CO_2, which result in Germany and therefore also in the EU by the German Unification. They were caused by significant production reduc-

tions in the former GDR and are often wrongly marked as gratis positions, although they are connected with tax money and investment of billions of Euro.

Umwelt- und Ressourcenökonomik
herausgegeben von Prof. Dr. Wolfgang
Pfaffenberger (Universität Oldenburg) und
Prof. Dr. Wolfgang Ströbele (Universität Münster)

Markus Utsch
Möglichkeiten und Grenzen einer internationalen Klimaschutzpolitik unter der Berücksichtigung des Verhaltens von Ressourcenanbieterstaaten
Bd. 1, 1994, 200 S., 22,90 €, br., ISBN 3-8258-2031-9

Dirk Hasse
Ökonomische Möglichkeiten und Grenzen des Einsatzes dezentraler Technologien in der Elektrizitätswirtschaft
Bd. 2, 1994, 216 S., 22,90 €, br., ISBN 3-8258-2044-0

Ingo Hensing
Terminmärkte als Form internationalen Rohstoffhandels – dargestellt am Beispiel von Mineralöl und Erdgas
Bd. 3, 1994, 248 S., 24,90 €, br., ISBN 3-8258-2046-7

Jürgen E. Blank
Marktstrukturen und Strategien auf dem Weltölmarkt – Spieltheoretische Betrachtungen
Bd. 4, 1994, 272 S., 24,90 €, br., ISBN 3-8258-2108-0

Cristina Chaves
Anreizfaktoren für die Einhaltung der Umweltnormen im Wasserbereich
Die Fallbeispiele Portugal und Bundesrepublik Deutschland im Kontext der EG-Integration
Bd. 5, 1994, 264 S., 24,90 €, br., ISBN 3-8258-2211-7

Nikolai E. Smirnov
Umweltpolitische Instrumente für die Erdölindustrie in Westsibirien
Bd. 6, 1995, 144 S., 20,90 €, br., ISBN 3-8258-2340-7

Gerhard Heese; Wolfgang Pfaffenberger; Stefanie Winkler
Chancen und Probleme einer Energie-Agentur im energiepolitischen Umfeld
Bericht über eine Begleituntersuchung zur Arbeit der Niedersächsischen Energie-Agentur
Bd. 7, 1995, 144 S., 19,90 €, br., ISBN 3-8258-2413-6

Chi-Tsung Lin
Die Energieversorgung Taiwans/R.O.C.
Struktur, Determinanten und Perspektiven als Beispiel für die energiewirtschaftliche Situation in südostasiatischen Schwellenländern
Bd. 8, 1995, 216 S., 24,90 €, br., ISBN 3-8258-2557-4

Petra Opitz; Wolfgang Pfaffenberger
Verpaßte Stunde Null?
Transformation am Beispiel der russischen Elektrizitätswirtschaft
Bd. 9, 1996, 136 S., 20,90 €, br., ISBN 3-8258-3046-2

Peter Jakubowski; Henning Tegner; Stefan Kotte
Strategien umweltpolitischer Zielfindung
Eine ökonomische Perspektive
Die Konkretisierung des Leitbilds "Nachhaltige Entwicklung" mit Hilfe von Umweltplänen steht ganz oben auf der umweltpolitischen Agenda. Die Wirtschaftswissenschaften bieten dabei bisher wenig Unterstützung. Daß die Ökonomie durchaus wichtige Beiträge zur Zielgestaltung leisten kann, zeigt diese Arbeit. Sie tritt allen Versuchen entgegen, sich bei der Zielsetzung auf Faustformeln oder Expertenurteile zu verlassen. Statt dessen wird ein Verfahren entworfen, das die Interessen aller Umweltnutzer berücksichtigt.
Das allgemeinverständlich angelegte Buch richtet sich an Studierende der Umweltwissenschaften, Planer, Politiker und alle, die sich in Theorie und Praxis mit umweltpolitischen Problemen befassen.
Bd. 10, 2., korr. Aufl. 1999, 192 S., 15,90 €, br., ISBN 3-8258-4112-x

Florian Baentsch
Umweltschutz im britischen Stromexperiment
Die umweltpolitischen Wirkungen der Strukturreform der britischen Elektrizitätswirtschaft hinsichtlich Schadstoffemissionen, Energieträgereinsatz und Energieeffizienz
Bd. 11, 1997, 352 S., 35,90 €, br., ISBN 3-8258-3307-0

Holger Schöttle
Analyse des Least-Cost Planning Ansatzes zur rationellen Nutzung elektrischer Energie
Wenn Endverbraucher aufgrund von Marktverzerrungen nicht immer von sich aus die energieeffizientesten Geräte auswählen, so kann durch Fördermaßnahmen die Wahl energieeffizienter Geräte unterstützt werden. Einen methodischen Rahmen hierfür bildet das Least-Cost Planning (LCP) bzw. Integrated Resource Planning (IRP) Konzept, das den Energieversorgungsunternehmen durch Investitionen in energiesparende Maßnahmen auf Kundenseite wirtschaftliche Strategien zur Beeinflussung der Nachfrage eröffnet.
Die vorliegende Arbeit befaßt sich daher mit der

LIT Verlag Münster – Hamburg – Berlin – London
Grevener Str./Fresnostr. 2 48159 Münster
Tel.: 0251 – 23 50 91 – Fax: 0251 – 23 19 72
e-Mail: vertrieb@lit-verlag.de – http://www.lit-verlag.de

Entwicklung einer modellgestützten LCP/IRP-Investitionsstrategie für ein Energieversorgungsunternehmen (EVU) und die Anwendung der Methode auf die Stadtwerke Rottweil. Hierbei wird in praxi untersucht, ob entsprechend dem Konzept von LCP/IRP neben erzeugungsseitigen Investitionen auch Fördermaßnahmen zum Energiesparen auf Seiten der Kunden aus gesamtwirtschaftlicher Sicht sinnvoll sind. Des weiteren werden Vorschläge zur Weiterentwicklung der energiewirtschaftlichen Rahmenbedingungen abgeleitet, unter denen ökologisch sinnvolle LCP/IRP-Konzepte auch aus der betriebswirtschaftlichen Sichtweise eines EVU attraktiv sind.
Bd. 12, 1998, 272 S., 24,90 €, br., ISBN 3-8258-4090-5

Hartmut Clausen
Rücknahmeverpflichtungen als Instrument von Abfallwirtschaftspolitik
Die hiesige Abfallwirtschaftspolitik wird zunehmend durch das umweltpolitische Leitbild einer nachhaltigen Entwicklung bestimmt. Eine stärkere Kreislaufführung von Produkten soll insbesondere durch Rücknahmeverpflichtungen erreicht werden. In der vorliegenden Studie wird analysiert, welche wohlfahrtstheoretischen Begründungen die Anwendung dieses speziellen Instruments rechtfertigen und unter welchen Bedingungen eine Schonung natürlicher Ressourcen sowie der Umwelt erreicht werden kann. Auch werden verschiedene Gestaltungsoptionen sowie wettbewerbspolitische Implikationen von Rücknahmeverpflichtungen diskutiert. In Fallstudien für die Bereiche Verpackungen, Altautos und Elektroschrott werden praktizierte bzw. geplante Entsorgungssysteme einer kritischen Analyse unterzogen und Verbesserungsmöglichkeiten aufgezeigt.
Bd. 13, 2000, 248 S., 20,90 €, br., ISBN 3-8258-4643-1

Peter Dworak
Technische und wirtschaftliche Möglichkeiten des Stromaustausches des russischen EES-Systemes mit dem skandinavischen NORDEL- und dem europäischen TESIS-System
Bd. 14, 2000, 136 S., 17,90 €, br., ISBN 3-8258-4885-x

Alexander Smajgl
Modellierung von Klimaschutzpolitik
Ein Allgemeines Gleichgewichtsmodell zur ökonomischen Analyse der Wirkungen von CO_2-Restriktionen auf den Einsatz fossiler Energieträger
Die Atmosphäre ist ein global öffentliches Gut und ihr Schutz stellt im politischen Prozess ein komplexes Problem dar. Daher bedarf es der ökonomischen Modellierung, um in Politikszenarien die potenziellen Effekte von CO_2-Restriktionen abzuschätzen.
Die vorliegende Arbeit spannt hierbei den Bogen von der Darstellung der Modellierungsmethodik bis hin zu der weltweiten Analyse gesamtwirtschaftlicher und sektoraler Wirkungen von Klimaschutzpolitik. Hierbei werden nicht nur Regionen mit CO_2-Restriktionen betrachtet sondern auch die regionenübergreifenden Effekte auf Förder- und Entwicklungsländer.
Bd. 15, 2002, 200 S., 30,90 €, br., ISBN 3-8258-5694-1

Uta Bahnsen
Zur Internalisierung grenzüberschreitender externer Effekte durch internationale Vereinbarungen, dargestellt am Beispiel des Übereinkommens zum Schutz der Meeresumwelt des Ostseegebietes von 1992 (Helsinki-Konvention)
Bd. 16, 2002, 216 S., 20,90 €, br., ISBN 3-8258-5760-3

Bernhard Hillebrand; Alexander Smajgl; Wolfgang Ströbele
Zertifikatehandel für CO_2-Emissionen auf dem Prüfstand
Ausgestaltungsprobleme des Vorschlags der EU für eine „Richtlinie zum Emissionshandel". Forschungsergebnisse einer Projektgruppe geleitet und herausgegeben von Bernd Heins
In der Klimaschutzpolitik steuert die EU-Kommission ein neues Instrument an, das helfen soll, die Kyoto-Ziele besser zu erreichen. Mit dem Richtlinien-Vorschlag vom 23.10.2001 wurde ein Konzept dazu konkretisiert: Verbindlicher Emissionshandel für Unternehmen mit bestimmten Anlagen soll ab 2005 eingeführt werden.
Die vorliegende Analyse dieser EU-Vorschläge zeigt auf: Der EU-Richtlinienvorschlag ist konzeptionell schwach, nicht abgestimmt mit den Kyoto-Instrumenten und von den Wirkungen her negativ für Ökonomie und Ökologie.
Bd. 17, 2002, 232 S., 24,90 €, br., ISBN 3-8258-6488-x

LIT Verlag Münster – Hamburg – Berlin – London
Grevener Str./Fresnostr. 2 48159 Münster
Tel.: 0251 – 23 50 91 – Fax: 0251 – 23 19 72
e-Mail: vertrieb@lit-verlag.de – http://www.lit-verlag.de